MINE HYDROLOGY

MINE HYDROLOGY

by

Roy E. Williams
Professor of Hydrogeology

Gerry V. Winter
Consulting Hydrogeologist

George L. Bloomsburg
Professor of Engineering Science

Dale R. Ralston
Professor of Hydrogeology

College of Mines and Earth Resources
University of Idaho
Moscow, Idaho

Published by the

Society of Mining Engineers, Inc.

Littleton, Colorado • 1986

Printed in the United States of America
by Port City Press, Inc., Baltimore, Maryland

Library of Congress Catalog Card Number 86-60202
ISBN 0-87335-056-1

Preface

This book presents an analysis of the state of the art methodologies and technologies available for predicting mine water inflow and its impact on the hydrology of the vicinity of the mine. Data collection techniques, equipment, and mathematical model simulation of mine water inflow are discussed and applied to two existing mines. Such analyses are important, not only because of impacts on water resources but also because underground mines frequently must deal with the hazards of ground water inflow both during the sinking of the shaft and during drift development. Inflow can be minor or it can occur in large quantities that can jeopardize an entire project. Obvious difficulties arise if large inflows of ground water occur to the underground workings. These problems frequently occur during initial shaft development when large inflows of ground water can severely hamper shaft sinking. Actions taken at this point are critical due to the limited time available to prevent flooding or partial flooding of the shaft. If the inflow has been predicted and is expected, the problem can be handled expeditiously. Ground water inflow to drifts can be a long-term expensive proposition due to the cost of pumping and the secondary effects of increased equipment wear particularly if the inflow is acidic. Ground water inflows into underground mines also present health and safety problems. These problems can range from the sudden catastrophic flooding of the mine to decreased wall stability and the inconveniences of working in continuously wet ground.

The rate of ground water inflow into both shafts and drifts affects both the primary mine pumping system design and the backup systems. Anticipated underground mine water inflow can be sufficient to warrant mine dewatering systems using pumping wells around the mine site. The proper design of a depressurization or dewatering well system for shaft and drift development requires both hydrogeologic information on the hydrostratigraphic units to be penetrated and the geometry of the boundary conditions at the site.

Impact on adjacent water resources is a major problem associated with ground water inflow to underground mines. This impact can range from merely depressurizing or dewatering the hydrostratigraphic units being mined to altering associated surface water bodies such as lakes, springs, and streams. Underground mines may affect hydrostratigraphic units that do not impact surface water systems due to the existence of confining units between the mine and the surface. Underground mines can alter both surface water systems and ground water flow systems when the systems interact hydraulically.

The evaluation of underground mine water inflow requires an understanding of the general hydrologic regime in the mine area. This understanding requires information on both the surface and ground water systems, depending upon the interaction of the two systems in the mine area.

Mine water inflow problems are associated with those hydrostratigraphic units that have the ability to discharge large volumes of water in a short length of time. This ability is dependent on the hydraulic conductivity of the unit and the fluid potential distribution (hydraulic head gradient) at the site. A unit with a high hydraulic conductivity and a low gradient can produce a high rate of inflow into a shaft or drift. Conversely, a unit with low hydraulic conductivity and a high gradient can also produce a high rate of discharge or inflow. A very high rate of inflow would occur if a high hydraulic conductivity is combined with a high hydraulic head gradient.

Boundary conditions influence significantly the rate of inflow into a shaft or drift. Recharge boundaries will maintain a higher rate of inflow with time than would be anticipated without boundaries. Barrier boundaries produce lower inflow rates.

Additional complexity is produced by the properties of heterogeneity and anisotropy. These properties can occur either in a simple single hydrostratigraphic unit system or in a multilayered hydrostratigraphic system. Further difficulties arise when fractures influence inflow into or among hydrostratigraphic units. Fracture spacing, dimension, and orientation play a significant role in the hydraulic properties of any fractured hydrostratigraphic system.

Data requirements for the analysis of mine water inflow must be based on an initial conceptual level of site investigation. First the area of investigation must be determined. This determination is based on existing geologic and hydrogeologic data. Springs and the gain-loss reaches of streams must be correlated with geology and potential recharge and discharge areas. This correlation constitutes the basis for the initial definition of hydrostratigraphic units and to some extent for the initial definition of boundaries. Ground water flow systems are defined in conceptual terms based on the definition of hydrostratigraphic units and boundaries. The mine plan is developed in part on the basis of this conceptual model of ground water flow systems. This procedure allows the investigating team and the mine planners to determine the areas requiring further study.

A second level of site investigation must follow. Spring and stream data must be collected both for baseflow characteristics and for long-term record keeping. Long-term records provide the basis for the determination of surface-ground water flow system interaction. Potentiometric data must be collected and assigned to individual hydrostratigraphic units. This procedure facilitates the determination of the distribution of both lateral and vertical fluid potential values. Hydraulic conductivity data should be collected from on-site pump tests. The storage characteristics of the hydrostratigraphic units must be determined and the conceptual model of ground water flow systems evaluated. Additional field data are collected as needed, depending upon this evaluation of the initial conceptual model.

Surface water system data collection utilizes conventional techniques such as precipitation monitoring and stream gaging. Stream gaging utilizes conventional equipment such as flumes, weirs, and current meters.

Geologic data can be collected for the ground water flow system evaluation using conventional geologic techniques. The conventional methods of test drilling provide information for the definition of the hydrostratigraphic units. Test drilling can provide both grab and core samples from the formations penetrated as well as information such as lost circulation during drilling. Hydrogeologic parameters such as fluid potential distribution can be obtained only through the installation of appropriate test wells. Hydraulic head distribution needs to be obtained in both lateral and vertical directions. Multiport piezometers are a recent innovation in this field that provide the investigator with a means of determining the vertical hydraulic gradient in multilayered hydrostratigraphic units through the use of a single borehole and casing. These piezometers employ multiple ports separated by borehole packers to isolate various hydrostratigraphic units or zones within a hydrostratigraphic unit.

Hydraulic conductivity values can be obtained for representative samples of the hydrostratigraphic units through laboratory analysis. These tests can consist of conventional constant head permeability or falling head permeability or more sophisticated radial permeability tests. These tests do not provide as accurate an evaluation of hydraulic conductivity as do field data from pumping tests.

Field data obtained from pumping tests constitute the best means of determining values of hydraulic conductivity and storage coefficient. Pump tests can be enhanced by the addition of multiple observation wells. Multiple wells also facilitate the determination of hydrogeologic boundary conditions in the vicinity of the mine.

Borehole geophysics is a technique not frequently used in the mining industry, but can provide information for the definition of hydrostratigraphic units. This technique can utilize existing or new boreholes in the area for the determination of unit boundaries based on geophysical properties. These properties can be determined through the use of caliper, resistivity, spontaneous potential, neutron-epithermal-neutron, and gamma-gamma density logs.

The data collected from laboratory and field tests at the mine site must be evaluated for the determination of hydrogeologic properties. Analysis of laboratory data is straightforward; however, the analysis of field data for hydrogeologic properties of transmissivity and storage coefficient is more subjective. A certain degree of experience is required in the evaluation of pump test data for the accurate determination of hydraulic properties and boundary conditions. Numerous techniques appear in existing literature for the evaluation of water table aquifers, artesian aquifers, artesian aquifers with leaky confining beds, and pump test data affected by boundary conditions. An accurate evaluation of field data is required for constructing models of ground water flow systems and the resultant calculations of shaft and drift inflow to a proposed mine.

A few techniques have been presented in the literature for predicting shaft and tunnel inflow. These techniques generally are based on simplifying assumptions such as line sinks and a constant water table or potentiometric surface. The mathematical model is a modern tool that is useful for the evaluation of changes within the hydrogeologic and physical system being investigated. These changes include both changes in water levels with time and in the dimensions of the proposed mine. The mathematical model facilitates repetitive and rapid calculations on complex systems that would require an enormous amount of time if hand calculation methods were attempted. In most cases the latter approach simply would not produce adequate results.

The requisite characteristics for mathematically modeling underground mine water inflow are that the model must be able to handle (1) a seepage surface, (2) perched water table, (3) heterogeneous materials, and (4) anisotropic materials. The mathematical model UNSAT2 and a few others met these criteria. UNSAT2 provides the investigator with a means of viewing the hydrogeologic system in three different modes. The first mode is axisymmetric in which the left boundary of a finite element mesh is used as the axis of symmetry. The second option is the horizontal planar mode. This mode allows the investigator to view the site as a single hydrostratigraphic unit in plan view. The third mode is the vertical planar mode. This mode simulates a vertical section cut through a portion of the site. All modes of operation can handle up to five distinct materials. These materials can exhibit variations in vertical and horizontal conductivity as well as conductivity at some angle of inclination to the vertical. UNSAT2 also has the capability of handling unsaturated flow either as an initial condition or as desaturation is developed as a unit becomes dewatered. We have selected UNSAT2 for use herein because of these versatilities.

We have developed a technique for using the UNSAT2 mathematical model to evaluate the inflow to a mine shaft using the axisymmetric mode of operation. This procedure allows the investigator to evaluate the shaft inflow as a function of depth. Shaft inflow can be determined based upon any rate of shaft sinking and any length and position of the open area of the shaft. Special conditions can include such symmetrical features as grout and symmetrically located dewatering wells. The influences of shaft diameter can also be evaluated with this procedure.

The same finite element mesh developed for the axisymmetric shaft inflow evaluation can be used for the evaluation of drift inflow where the drifts occur within the same hydrostratigraphic units previously investigated. Drift inflow is evaluated through the use of the vertical planar mode of operation. This technique involves a dimensionless evaluation of inflow to a drift as the drift extends in a straight line perpendicular to the shaft.

Certain limitations are imposed on the use of these techniques for evaluating shaft and drift inflow. The first limitation is on the number of materials as described previously. The second limitation is the restriction placed by the mode of operation in use. The axisymmetric mode of operation requires that boundary conditions be simulated in a radially symmetric manner. This restriction is not significant where boundary conditions can be considered infinite in areal extent. The environmental impact of a potential mining operation can be evaluated in part by using this mode of operation where the surface formation can be included in the five material layer requirements as noted previously.

The horizontal mode of operation permits the evaluation of complex boundary conditions by modeling the mine site in plan view. Depressurization or dewatering wells can be simulated by assigning a rate of discharge at a node. Mining occurs by methodically "removing" blocks of material that have been mined out. Mining is simulated by changing the material code on those elements "removed" to a fictitious material with high hydraulic conductivity and porosity. A constant potential of zero is assigned to those nodes at the corners of the element "removed." Successive elements are removed as mining progresses.

The degree of sophistication required for evaluating mine water inflow should be based upon the requirements of the mine planner and operator. In general a "ball park" approach to calculating mine water inflow can be obtained using existing hand calculation techniques. A more sophisticated approach utilizes the techniques presented herein. A still more sophisticated evaluation requires an even more complex mathematical model than the one presented. Such models (three-dimensional as well as two-dimensional) are available but costs of data acquisition and model use rise accordingly.

Uncertainties and risks are inherent in any simulation of mine water inflow. The major uncertainties are the location and presence of boundaries at depth. Unknown recharge boundaries or boundaries exhibiting a high hydraulic conductivity are particularly dangerous to an evaluation. The presence of unknown fractures and faults is one of the major problems in simulating or developing a mine. Vertical fractures and faults often are difficult to locate utilizing surface methods of detection. Underground methods such as drilling ahead of the drift usually intercept such features. These features can constitute obvious mine water inflow hazards if encountered during mining. Such discontinuities can provide a means of transmitting large volumes of water in a short length of time. A fault can provide a means of recharge to the mine from adjacent hydrostratigraphic units or a fault can act as a semipermeable or completely impermeable barrier boundary.

We have applied the techniques presented to the evaluation of mine water inflow at an existing uranium mine located in New Mexico. This application illustrates our methodology for predicting inflow to a shaft as it progresses downward. The mine in question has hydraulic conductivity and storage characteristic values defined for three hydrostratigraphic units located at the bottom of the shaft. Potential mine water inflow problems had been anticipated by the mining company prior to shaft sinking by conventional techniques.

Our evaluation of mine water inflow to the shaft was complicated by several factors. A barrier boundary is located 2000 to 3000 ft (610 to 904 m) from the shaft. This boundary is semipermeable as opposed to an ideal barrier boundary. The existence of numerous tests without complete records of the amount of water removed prior to the initiation of data collection constituted an additional complicating factor in our analysis. The absence of these data complicated our simulation of the pretest conditions. A ring of depressurization wells around the shaft also complicated our simulation. These wells were not pumped at a continuous rate but were pumped at rates determined by the potentiometric level in the vicinity of the shaft. Grouting in the uppermost hydrostratigraphic unit that the shaft penetrated also had to be simulated.

Shaft inflow was modeled using a constant shaft sinking rate combined with an open drill hole. The open drill hole was used at the bottom of the shaft to simulate grouting operations in the upper hydrostratigraphic unit. We simulated the unlined portion of the shaft with a seepage face set at atmospheric pressure. The open portion of the shaft was moved downward as it moved toward the first drift. The drift was left open and the shaft was allowed to discharge water for a period of time estimated to be used by the mining company for initial drift work at this level. This face remained open while shaft sinking continued. Two seepage faces (first drift and shaft) existed during this phase of simulation. The shaft was then sunk to the second level of drifts. Mine inflow was principally shaft inflow up to this point. The vertical mode of UNSAT2 was used with the same mesh to determine the rate of inflow into each drift. Each drift level was evaluated separately and the results of the upper and lower drifts and the shaft were added for the total mine water inflow. Simulated inflow rates are within about 10% of measured inflow rates after the inflow rate exceeds 300 gpm (1.1 m^3/min). Increasing deviation, after 390 days, of our simulated data from measured inflow rates can be attributed to our assumption that the drift extends in a linear fashion from the shaft. In reality drifts do not proceed in this fashion but form a complex interlocking pattern that would require a three-dimensional simulation.

Inflow to a lead mine located in Missouri was simulated by extraction of mine blocks in the finite element mesh. The mine consists of a vertical shaft with a room-and-pillar mine at the bottom. The horizontal mode of UNSAT2 was used for this simulation. This simulation was complicated by the presence of numerous and irregular shaped boundaries that are semipermeable in nature. In addition, only limited data were available on the column of hydrostratigraphic units at this site. Consequently we were forced to evaluate this mine based on a single layer model consisting of two distinct hydrostratigraphic units. The difference in hydraulic conductivity between the two units is approximately one order of magnitude. Our simulation of this mine was complicated also by the pumping of dewatering wells, without adequate record keeping, prior to shaft sinking. We simulated inflow into the lead mine by removing blocks (elements) of material from the mined-out area. This removal was accomplished by setting a constant potential on the nodes of the finite element mesh that formed the boundaries of the block removed and assigning a high hydraulic conductivity value to that block of material. Mining was simulated in a clockwise fashion as sequential blocks of material were removed. Each block was replaced by a block of material having a high hydraulic conductivity at a constant potential of zero. Inflow simulated in this fashion was not represented by a smooth curve of discharge as would be expected in a real case. Instead an initial high rate of inflow occurred when a block was first removed. As time progressed the rate of inflow dropped accordingly to a point lower than would be anticipated in an actual mine. Actual mine inflow would best be simulated by an average of the high and low values using this method of simulation. The average simulated inflow was 660 gpm (2.5 m^3/min) vs. a reported 600 to 800 gpm (2.3 to 3 m^3/min).

Subsequent chapters present details of data collection techniques and data needed for predicting mine water inflow. Details of the application of the finite element model for this purpose also are presented. We believe that these procedures can be used for instruction and as guidelines in the treatment of problems in mine hydrology.

The Authors
University of Idaho

Acknowledgments

This manuscript was prepared in large part under the support of the US Bureau of Mines, Contract number J0100013. The project was initiated under the USBM Conservation and Development Program. We are grateful for the assistance of Bill M. Stewart of the Spokane Research Center. We appreciate also the assistance of Dan Kealy and Frank M. Naughton, also of the USBM. We express thanks to Sue Line and Andrea Beckett for their assistance in handling many of the logistical problems that are inherent in a multiple authorship effort of this type. This manuscript is a summary of work completed during the period Mar. 3, 1980 to Aug. 3, 1982.

Table of Contents

1

Introduction

It is imperative that the hydrogeologic framework and the ground water flow systems of a potential mine area be investigated prior to initiation of mining. Such investigations are necessary so that potential mine water inflow problems can be anticipated and controlled. Obvious difficulties can arise if large inflows of ground water occur in underground workings. Costs of water removal and disposal, if not foreseen and included in economic planning, can severely limit or eliminate profit from a mining operation. Safety problems associated with large inflows are an additional potential problem.

The inflow of water into a mine necessarily alters the associated ground water system. The inflow also may impact nearby surface water regimes, depending upon the interaction of the surface water and ground water systems. The inflow of water into a mine inevitably causes a decline in the ground water levels in the vicinity of the mine. This decline may be restricted to a small area or it may extend a considerable distance, depending upon the hydrogeologic properties of the associated hydrostratigraphic units. A decrease in surface flow will result if the area of ground water level decline intersects a stream or spring that is hydraulically connected to the ground water system.

Mine inflow problems historically have been evaluated after-the-fact. Premining hydrologic investigations only recently have been initiated. Brown's (1977)* report on anticipated hydrologic impacts of oil shale mining in the Piceance Basin of Colorado is an example of premining hydrologic planning. Observed difficulties inherent in designing a data collection network and then utilizing the data to predict mine inflow have led us to the preparation of this manuscript.

This manuscript reflects a two-phase effort. The first phase describes and develops analytical and numerical techniques for mine inflow prediction. Emphasis in this section is placed on the selection and application of data collection techniques and data analysis procedures. The second phase applies the prediction technique(s) developed during phase 1 to two existing mines. The objective of phase 1 efforts is to evaluate the hydrologic variables associated with surface water systems and ground water systems that can affect the inflow of water into an underground mine. Subsequently these variables are incorporated into predictive techniques for estimating underground mine water inflow and its effect on associated ground water and surface water systems.

The specific objectives of phases 1 and 2 are to:

1) Determine the hydrologic parameters required for predicting underground mine water inflow and the potential effects of the inflow on surface and ground water systems. The hydrologic parameters represent both premining and post initiation conditions.

2) Outline methods for obtaining the necessary input data for the predictive techniques in premining and post initiation periods.

3) Present existing numerical techniques for predicting underground mine water inflow including assumptions and required input.

4) Evaluate existing numerical techniques previously tabulated based on their applicability to mine water inflow prediction.

5) Attempt to develop analytical techniques for predicting underground mine water inflow.

6) Develop, through modification of existing models, a general mathematical model as a numerical technique for predicting underground mine water inflow and the resultant effects on surface and ground water systems.

7) Apply the predictive techniques to two existing mines selected for this purpose.

8) Compare the predicted inflow rates to measured inflow rates at both mines.

9) Compare predicted impact on district hydrology to observed impacts where possible.

10) Evaluate the predictive technique.

* All references are given in Chapter 9, References.

2

Underground Mine Water Inflow

GENERAL HYDROLOGIC REGIME

The total hydrologic cycle must be considered when evaluating potential ground water problems in a mining environment. Precipitation is the ultimate source of all surface water and essentially all ground water.

Stream flow during low flow periods often originates as seeps or stream gains (base flow) where the water table intersects the land surface. A stream that is higher in elevation than the ground water system may lose flow, thereby recharging the ground water system. The magnitude of stream gain or loss is dependent upon the hydraulic conductivity and related properties of the underlying hydrostratigraphic units. Ground water moves constantly from recharge areas to discharge areas (streams and springs) usually at very slow rates.

Large quantities of ground water may be intercepted in many underground mining environments. Construction of an underground mine in saturated formations may encounter ground water inflows ranging from very small to very large and from continuous to intermittent depending upon the hydraulic properties of the hydrostratigraphic units penetrated.

SURFACE WATER VARIABLES

Precipitation occurs in different forms, varying amounts, and with a wide assortment of temporal and areal distributions in the various mining districts within the United States. Mine inflow can be affected in a matter of hours by precipitation events where there is a direct hydraulic interconnection between the mine and the surface through fractures or a mine opening. The timing, form, amount, and distribution of precipitation should be evaluated in a mine inflow study.

A portion of the precipitation that falls in a topographic basin moves as sheet flow to runoff channels (Fig. 1). Some of the precipitation infiltrates the unsaturated soil zone where it can either move laterally as interflow to the surface system to augment stream flow or move downward as recharge to the ground water system. The unsaturated zone (called the zone of aeration or vadose zone) can range in thickness from essentially zero to a few hundred feet. The saturated zone or ground water system occurs below the water table (zone where hydrostatic pressure is greater than atmospheric pressure) where all the pores are filled with water (Freeze and Cherry, 1979, p. 44). Moisture can be lost from any of the systems (surface water system, unsaturated zone, and ground water system) by the processes of evaporation and transpiration which return moisture to the atmosphere.

Stream flow consists of runoff and interflow components plus an additional base flow component (Fig. 2). Base flow constitutes ground water discharge to a stream. Base flow plays an increasingly significant role as total stream flow declines. Base flow is important because it constitutes much if not all stream flow between periods of precipitation when runoff and interflow have largely dissipated.

INTERACTION OF SURFACE WATER AND GROUND WATER SYSTEMS

Recharge from a surface water system to ground water systems can occur via unsaturated or saturated flow. The amount of recharge is dependent primarily upon the vertical hydraulic conductivity of the formations underlying the stream and the availability of water. Streams that flow over low hydraulic conductivity formations lose little water even if the ground water levels are significantly lower than the stream. A ground water system may contribute flow to a surface water system but normally only by saturated flow. This contribution usually takes the form of seeps, springs, and the base flow of streams.

GROUND WATER VARIABLES

Introduction

The variables that affect ground water flow are discussed in increasingly complex theoretical situ-

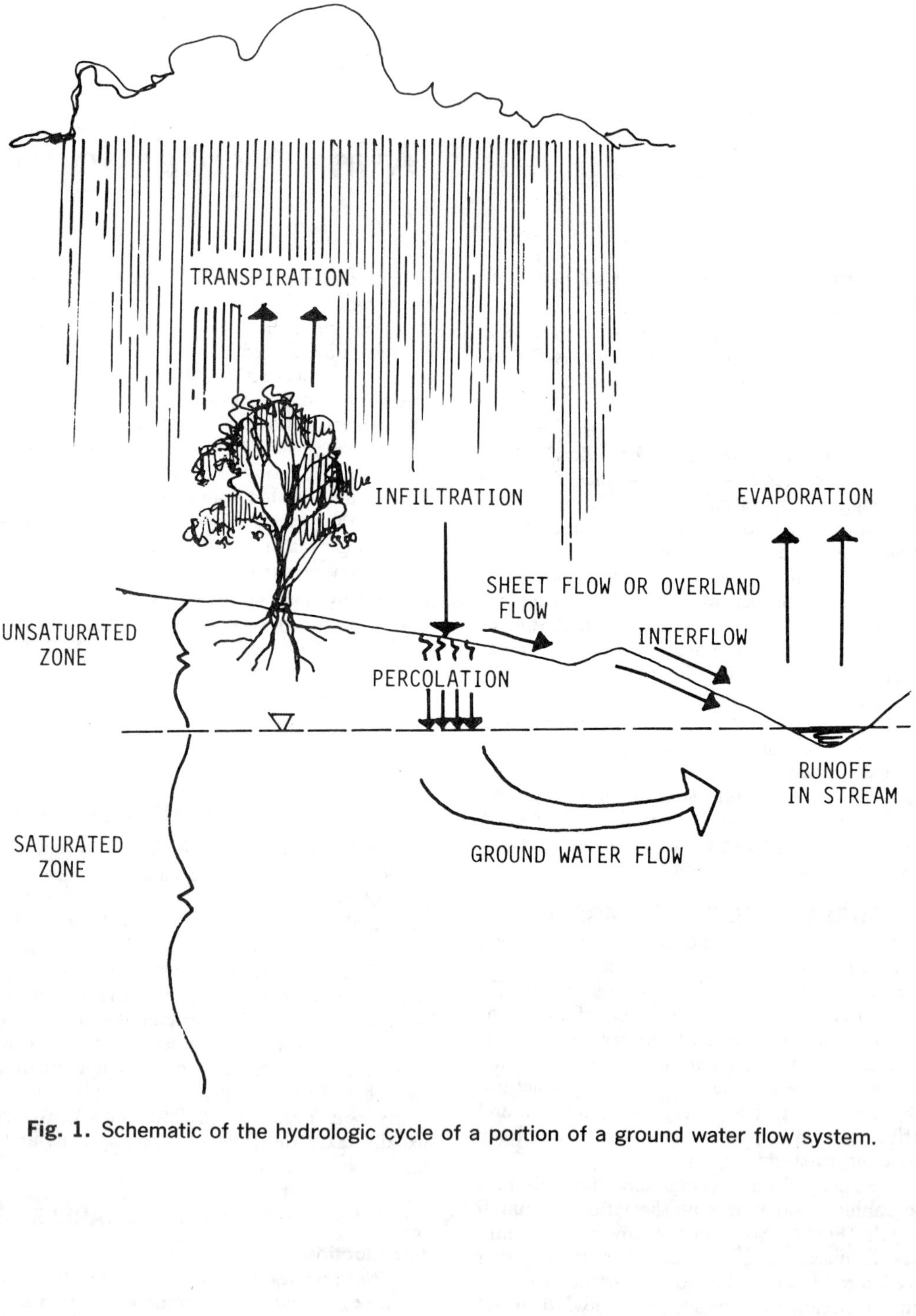

Fig. 1. Schematic of the hydrologic cycle of a portion of a ground water flow system.

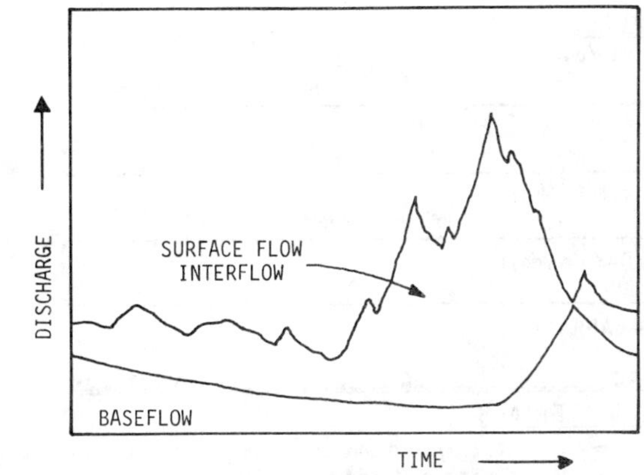

Fig. 2. Typical stream hydrograph components, modified after Domenico (1972, p. 49).

ations in the following sections. The initial discussion describes the basic components that control the flow of ground water in a simple, single layer system. Discussion of more complex multilayered systems follows.

Simple Single Layer System

The most simple hydrologic setting for an underground mine is a single layer system consisting of a single hydrostratigraphic unit (Fig. 3). A hydrostratigraphic unit has areally consistent hydrogeologic properties and defined boundaries. The boundaries of a hydrostratigraphic unit may or may not coincide with the conventional geologic boundaries defined by lithology, mineralogy, or other properties.

The potentiometric surface is a measure of the energy level in a hydrostratigraphic unit; it is determined by the elevation of the static water level in cased wells open only to the hydrostratigraphic unit. It is usually expressed as an elevation or height above some datum, typically sea level. Since the potentiometric surface is an energy level, water flows from a higher to a lower potential and hence from a higher to a lower ground water level. Contour maps of water level elevations (a reflection of energy levels) are used to determine directions and characteristics of ground water flow. Ground water flow through geologic material is governed by a simple relationship (Darcy's law) that describes the loss of energy as a fluid moves through porous media:

$$q = \frac{Q}{A} = -K\frac{dh}{dl} \tag{1}$$

where q is the specific discharge in units or dimensions of LT^{-1}, Q is the flow rate (L^3T^{-1}), A is the cross-sectional area (L^2), K is the hydraulic conductivity (LT^{-1}), dh is the difference in static head between two points (L), dl is the distance between those two points (L), and dh/dl is the hydraulic gradient (Lohman, 1972, p. 10). The units indicated are valid for any consistent set of (L) length-time (T) units such as feet and days or meters and seconds. This format is used herein except as otherwise noted.

Darcy's law is valid so long as the flow is laminar. Some evidence suggests that an upper limit exists for Darcy's law where flow velocities are sufficiently high that turbulent and/or nonlinear flow occurs. The law also may not be applicable for extremely fine grained materials under very low gradients. Darcy's law is valid, however, for the vast majority of ground water flow situations found at mining sites. It is assumed to be valid throughout this book.

Hydraulic conductivity is a measure of the ability of a geologic material to transmit water. It is dependent on the amount of pore space present in the medium (porosity), the degree of interconnection of the pore space, and the geometry and size of the pore space. Fig. 4 gives approximate values of (K) for a wide range of geologic materials. Hydraulic conductivity is a function of both the porous media and the fluid. Fluid properties, however, can be considered constant for most mining situations.

Most geologic materials exhibit some form of planar feature such as bedding or fracture planes. Hydraulic conductivity parallel to these planar features usually is greater than the hydraulic conductivity perpendicular to these planar features. This property is referred to as anisotropy. The anisotropy near an underground opening will have a major impact on the quantity of water entering the mine, as will be shown subsequently.

The properties of hydraulic conductivity are seldom constant over any given large area. This variation is referred to as heterogeneity. A hydrostratigraphic unit is homogeneous if a hydraulic property is independent of its position (Freeze and Cherry, 1979, p. 30). The location and temporal characteristics of mine water inflow are dependent on whether the geologic material has uniform hydraulic properties or properties that vary widely. The heterogeneous, anisotropic distribution of hydraulic conductivity near most mine sites requires that considerable field investigation be completed

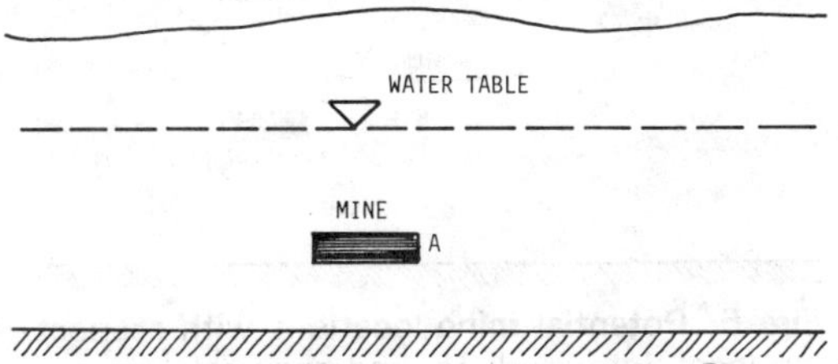

Fig. 3. Simple single layer system.

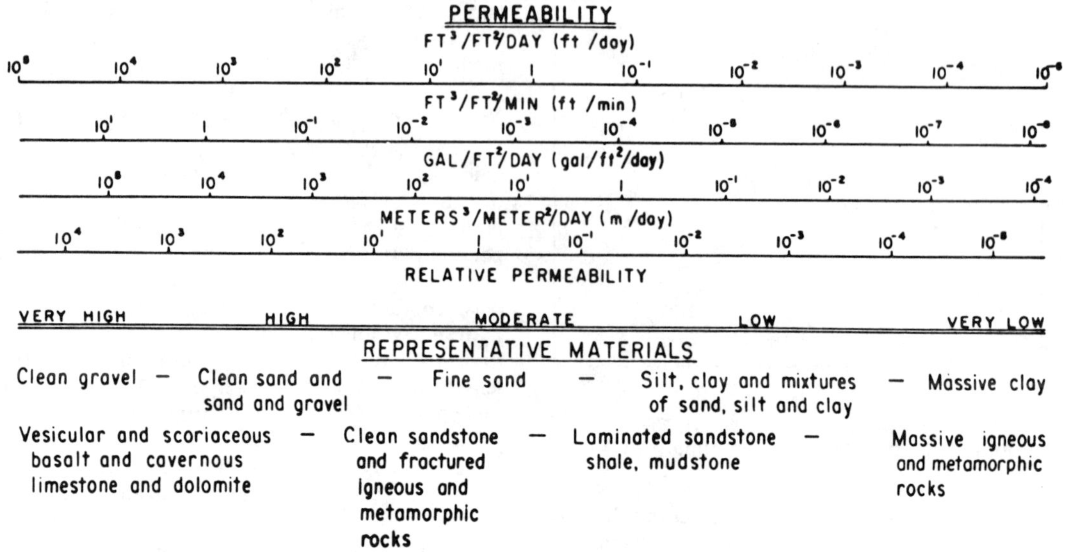

Fig. 4. Comparison of hydraulic conductivity (permeability) and representative materials (Verner, et al., 1974, p. 29).

before the ground water flow systems can be evaluated.

The quantity of water a single hydrostratigraphic unit may take in or release from storage is called specific yield or unconfined storage coefficient. Specific yield is the volume of water released from storage via gravity drainage with a unit decline in hydraulic head per unit surface area of the hydrostratigraphic unit (Freeze and Cherry, 1979, p. 61). The volume of water released is due to the drainable porosity and hence the amount of interconnected pore space.

A ground water flow system includes the entire set of pathways that water follows through the ground between hydraulic boundaries. It includes the recharge area, lateral flow area, and discharge area. The flow pattern is controlled by hydraulic boundaries, by the quantity and distribution of recharge, topography, and the spatial distribution of hydraulic conductivity. Fig. 5 illustrates one combination of these factors. Flow lines are shown on the diagram to illustrate the path a unit mass of water would follow from a specific recharge site.

Ground water flow systems are designated local, intermediate, or regional. These designations aid in categorizing field data and in assessing potential mining impacts. Toth (1963) defined these systems as follows:

1) A local system has its recharge and discharge areas adjacent to each other at a topographic high and low, respectively.

2) An intermediate system will have one or more topographic highs and lows located between its recharge and discharge areas that do not occupy the highest and lowest elevations in a basin.

3) A regional system has its recharge and discharge areas occupying the highest and lowest elevations, respectively, in a basin.

The patterns of ground water flow to a large extent control concentrations of dissolved constituents and temperature of the ground water. The mineralization of ground water is dependent on the time of contact of the water with soluble minerals in a porous medium and thus on velocity of movement and length of flow path. Higher concentrations of dissolved solids in ground water normally occur along longer flow paths where velocities are low. Seasonal variations in air temperature are usually damped out at a formation depth of about 9 m (30 ft) (Domenico, 1972, p. 281). Ground water

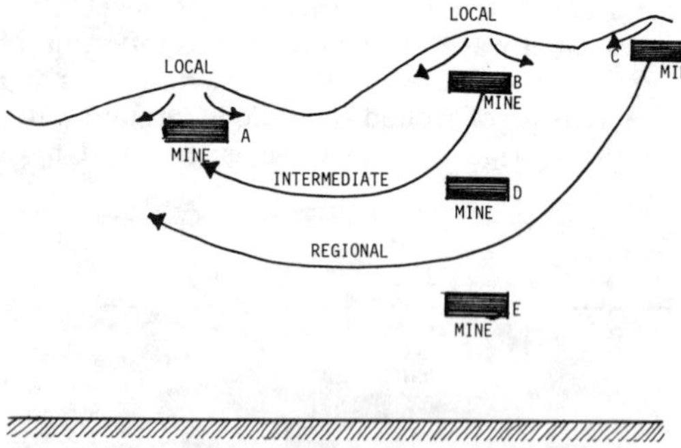

Fig. 5. Potential mine locations with respect to local, intermediate, and regional ground water flow systems.

at that depth has a nearly constant temperature that approximates the mean annual air temperature of the region. Below that depth, the water temperature is controlled by the flow of heat from the interior of the earth. In some cases, ground water temperatures can be used to estimate the depth of circulation of ground water flow systems. A spring that has a temperature near the mean annual air temperature probably issues from a shallow local flow system. High ground water temperatures usually indicate the presence of deep, regional flow systems (Domenico, 1972, pp. 281–282). These easily obtained data can be useful in the early stages of mine planning. The data can be collected during exploration drilling.

Multilayered System

A multilayered system is defined as a ground water flow system or systems consisting of more than one hydrostratigraphic unit. The differentiation of individual hydrostratigraphic units is based on their respective hydrogeologic properties. Ideally, an individual unit would have a uniform hydraulic conductivity at all points in the unit. In reality, hydrostratigraphic units are differentiated by greater changes between units than within units. Designation of hydrostratigraphic units is an important issue in itself, as will be discussed subsequently.

A multilayer system frequently is defined by hydraulic properties that coincide with geologic delineations. Changes in depositional environment that lead to formational definitions are often significant from a hydrogeologic point of view. Figs. 6 and 7 graphically illustrate the need to determine the attitude of each hydrostratigraphic unit and the elevation of the top and the bottom of each hydrostratigraphic unit. The structural setting of a

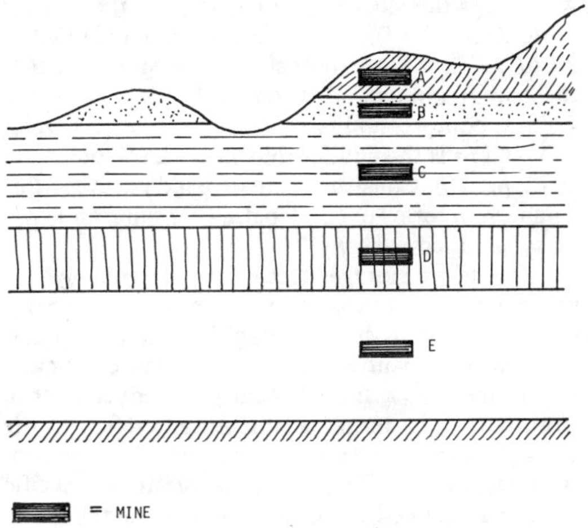

Fig. 6. Potential mine locations in a horizontally stratified, multilayer system.

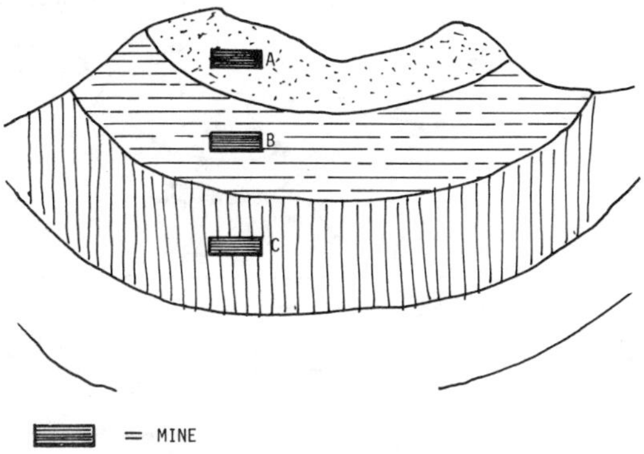

Fig. 7. Potential mine locations in a synclinal multilayer system.

complex multilayer system can control the pattern of ground water flow in the vicinity of a mine because the structural setting can control the distribution of hydraulic conductivity.

The hydraulic conductivity by definition should differ significantly from one hydrostratigraphic unit to another in a multilayered system. The differences in hydraulic conductivity among hydrostratigraphic units is especially critical to the formulation of conceptual models of ground water flow systems and ultimately the simulation of mine water inflow. Freeze (1967) first demonstrated the effect caused by altering the location of hydrostratigraphic units with significantly different hydraulic conductivities. Fig. 8 illustrates one example of the alteration of ground water flow systems by the presence of a unit with a high hydraulic conductivity. A second discharge area occurs near the down gradient end of the high hydraulic conductivity unit.

The fluid potential distribution (head distribution) within each hydrostratigraphic unit must be defined to predict inflow to an underground mine. Fig. 8 illustrates the significant alteration of the equipotential lines (lines connecting points of equal potential or hydraulic head) from 8a to 8b by the presence of a unit with a high hydraulic conductivity. The potential distribution in both the horizontal and vertical directions is important in defining the spatial direction of ground water movement. Vertical movement of ground water will occur when there is a different potentiometric level in adjacent hydrostratigraphic units. A higher potential in the upper unit indicates that there is downward movement of ground water. The lateral or horizontal movement of ground water within a hydrostratigraphic unit is caused by a higher potential at one point than another point within the same unit. Flow is always from a higher to lower potential (from the right to the left in Fig. 8a). A mine introduced into a ground water flow system acts as a potential energy sink that causes water to

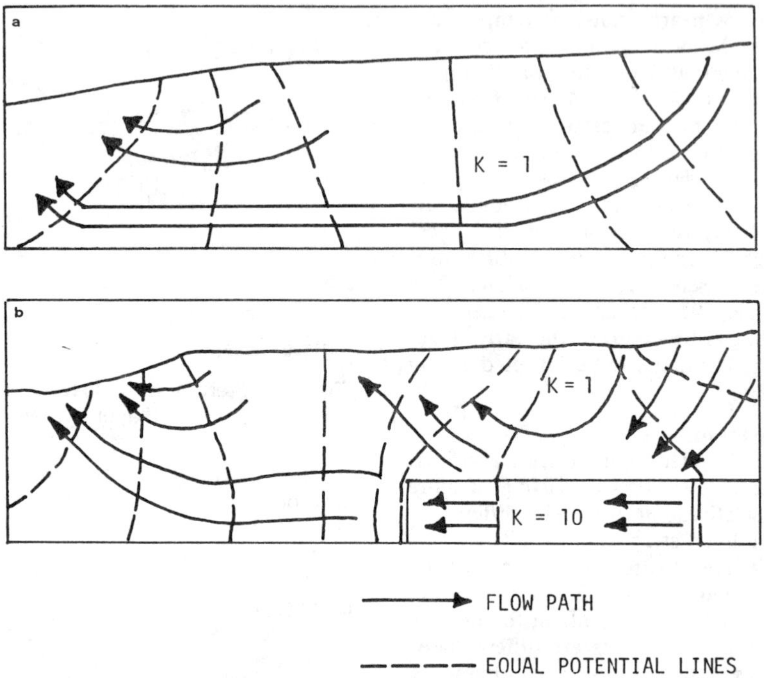

Fig. 8. Alteration of ground water flow systems by the presence of a high hydraulic conductivity unit (modified after Freeze and Witherspoon, 1967). (a) Uniform hydraulic conductivity. (b) Hydrostratigraphic unit with low hydraulic conductivity ($K = 1$) or high hydraulic conductivity ($K = 10$).

move toward it from adjacent zones of higher potential.

Two different concepts of storage coefficient (confined and unconfined) are necessary to define the quantity of water released from storage when a mine creates a reduction in potential in a multilayered system. The water in the uppermost layer is subject to gravity drainage; therefore, the unconfined storage coefficient for this zone is specific yield. Water is derived from storage from the lower layers with a change in potential even though none of the materials are desaturated. This concept is based on a confined storage coefficient. A decrease in potential in zone E of Fig. 6 would cause a slight expansion in the water in that layer (water is compressible—a decrease in potential is accomplished by a decrease in fluid pressure). The decrease in fluid pressure in the aquifer would transfer some of the weight of the overlying material to the porous medium (the effective stress or grain-to-grain contact in the porous medium and the fluid pressure combine to support the overlying weight). The additional weight on the aquifer skeleton will cause some compaction. This causes water to be released from storage in the lower layers or confined aquifers by a combination of expansion of water and compaction of the porous media. Obviously, the quantity of water released by these processes is small. The equation describing the storage coefficient of a confined aquifer is (Domenico, 1972, p. 220):

$$S = m\rho_\omega g(\propto + n\beta) \ , \qquad (2)$$

where S is the storage coefficient, ρ_ω is the density of fluid (M/LT^2), g is the acceleration due to gravity (L/T^2), n is the porosity (dimensionless), β is the compressibility of water (LT^2/M), $\propto$ is the vertical compressibility of element skeleton (LT^2/M), and m is the aquifer thickness (L) (Domenico, 1972, p. 220). Consistent units are indicated for this equation in terms of (L) length, (T) time, and (M) mass as noted earlier.

The term storage coefficient is probably the prevalent term used to describe the storage property of a confined hydrostratigraphic unit or aquifer. Storage coefficient is synonymous with coefficient of storage and a more recently introduced term, storativity. The term specific storage refers to the storage property of a unit thickness of the hydrostratigraphic unit or aquifer in question. Specific storage is defined as $\rho_\omega g(\propto + n\beta)$. A thorough discussion of the concepts used to define the storage properties in the fields of hydrogeology, petroleum engineering, and soil mechanics is presented by Narasimhan (1980).

The differentiation between storativity for confined or unconfined aquifers is based upon the potentiometric level for the hydrostratigraphic unit in question. Specific yield (gravity drainage) occurs in those cases where the potentiometric level is below the top of the hydrostratigraphic unit. Fig. 9 shows unit 1 as an unconfined aquifer. A confined system exists when the potentiometric level for a unit, such as unit 3 in Fig. 9, is above the contact between the unit and the overlying unit. Such a situation might occur if units 1 and 3 are sands and unit 2 is a clay. The storage coefficient of unit 3 becomes specific yield during the development of a mine if pumping causes the potentiometric level for that zone to drop below that contact with the overlying unit. Such a change in potential would occur if a mine were constructed in unit 3 of Fig. 10. The key to defining the storage coefficient of a particular unit is the position of the potentiometric level with respect to the upper contact of the unit as shown in Fig. 9. One unit can have both storage characteristics at the same time but at different locations (Fig. 10). Proper identification of the storage characteristics is essential for accurately determining the rate and duration of mine water inflow as a mine expands.

Ground water flow in single layer or multilayer systems is controlled by the location and properties of the hydrostratigraphic units, the location of recharge and discharge areas, magnitude and locations of the recharge and discharge, and the location of barrier boundaries. Barrier boundaries are units that have low hydraulic conductivity which will

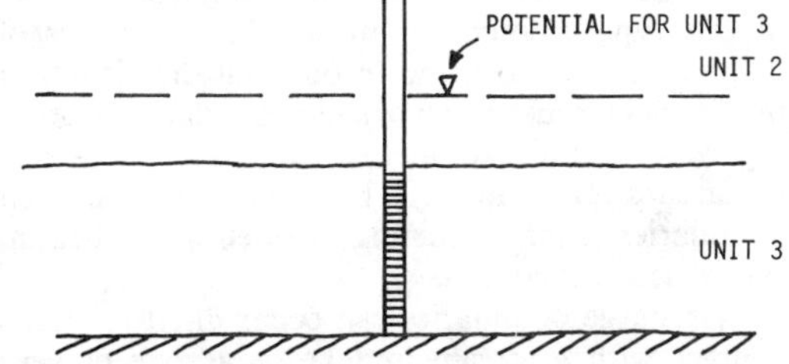

Fig. 9. Distinction between confined and unconfined hydrostratigraphic units based on potential (ground water levels).

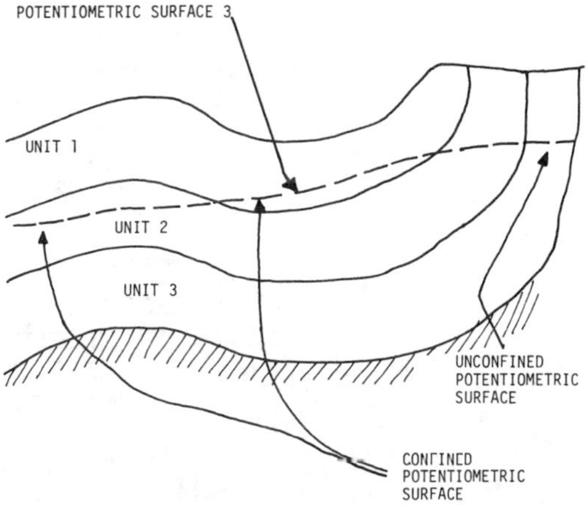

Fig. 10. Relationship of confined-unconfined potentiometric surface in a hydrostratigraphic unit.

retard ground water flow although they may contain large amounts of water in storage. A unit with a high hydraulic conductivity can have large quantities of water in storage and support a very significant ground water flow system. A flow system can exist only if there is recharge to the unit, a continuous flow path, and discharge from the unit. The horizontal potentiometric surface in Fig. 3 indicates that, at least prior to mining, there is no lateral movement of ground water in that situation. A sloping potentiometric surface must exist as illustrated in Fig. 10 before flow can occur in a lateral direction.

Geologic structure(s) can greatly affect the path that flow systems follow by altering the distribution of hydraulic conductivity. Fig. 11 illustrates how the direction of ground water flow can be controlled by geologic structure. The shallow flow system in unit 1 discharges at the topographic low along the axis of the syncline. Ground water in unit 4 moves from a recharge area in the valley on the right under the high ridge to discharge in the valley on the left side of the figure.

Fig. 11 illustrates the types of ground water flow systems that can exist. Hydrostratigraphic units 2 and 3 support only shallow ground water flow near the axis of the anticline. These low hydraulic conductivity units support more flow at this area because of the weathering of a fractured zone since more fracturing can be expected near the axis of an anticline or syncline. Unit 1 has a high hydraulic conductivity and supports only local flow systems due to the influence of topography. Unit 4 supports an intermediate flow system due to its high hydraulic conductivity and the influences of topography. The barrier boundary created by the fault is effective in blocking flow. The fault forces

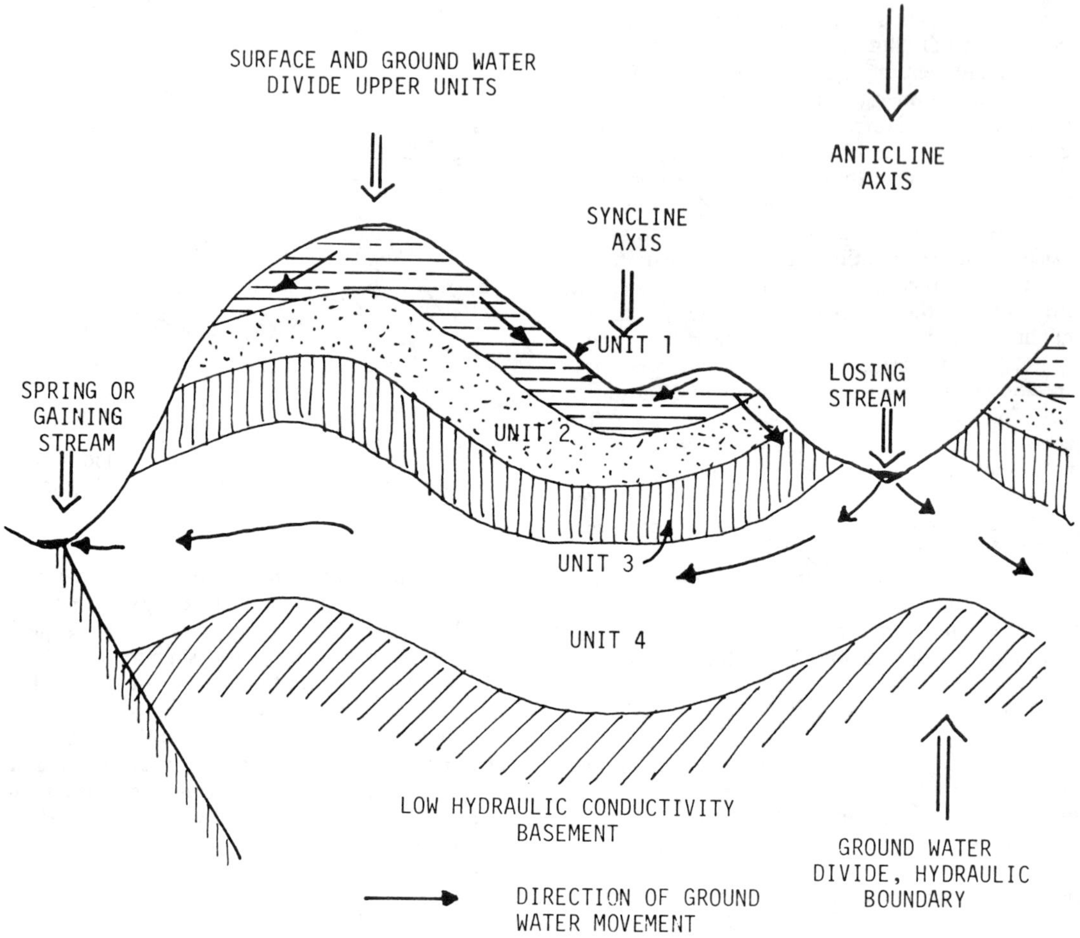

Fig. 11. Geologic influence on ground water flow systems.

the ground water flow in unit 4 to the surface where it discharges as a spring or a gaining reach of a stream. Conversely, a losing reach of the stream exists above the axis of the anticline.

Physical (geologic) boundaries in ground water flow systems exist where there is a lateral discontinuity in hydraulic conductivity. A significant decrease in the hydraulic conductivity creates a barrier boundary across which little movement of ground water can occur. Recharge boundaries are created where the aquifer unit intercepts surface water systems or geologic units with significantly higher storage capacity and hydraulic conductivity than the aquifer in question.

Barrier boundaries can exist in numerous forms. The fault noted in Fig. 11 is one type of barrier boundary that can block ground water flow. Unit 4 has been displaced and placed opposite the low hydraulic conductivity basement material. Faults can act as barriers blocking the flow of ground water in materials such as basalt (Noorishad, et al., 1971, pp. A4–A5). The result is a high potentiometric level on the up gradient side of the fault with a much lower potentiometric level on the down gradient side. Fault gouge is a primary cause of low hydraulic conductivity in fault zones, but this is very dependent upon the type of material involved and its thickness. Hydrogeologic boundaries of this type account for most of the inrushes of water that have occurred throughout the history of mining. The penetration of such boundaries by mining can be a threat to safety as well as to the overall economics of a mine. The delineation of such boundaries prior to mining is essential to health, safety, and economics.

Recharge boundaries can occur due to a unit's contact with a stream or lake or a zone of very high hydraulic conductivity. The water body can fully penetrate the unit of interest as shown in Fig. 12a or only partially penetrate the unit as in Fig.

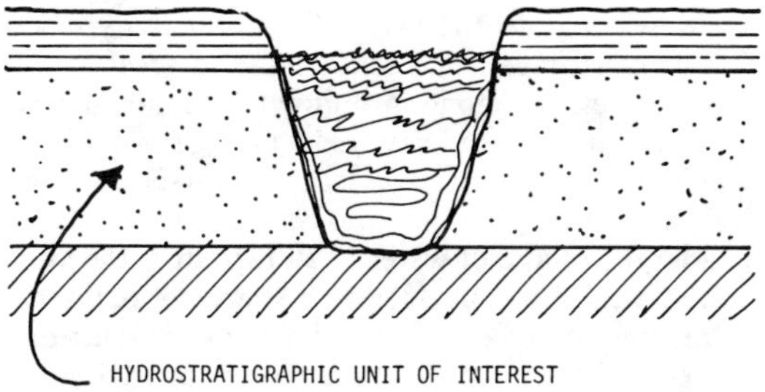

(a) Water body fully penetrates unit

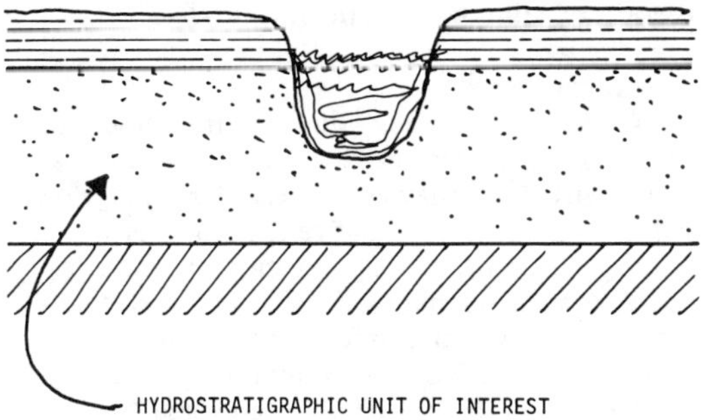

(b) Water body partially penetrates unit

Fig. 12. Surface water bodies as recharge boundaries.

12b. Recharge boundaries can consist of saturated zones of very high hydraulic conductivity. An example of this type of boundary is a highly brecciated fault zone as illustrated in Fig. 13. Water stored in the extensive fault zone in Fig. 13 is not confined so a considerable amount of water can be transmitted to unit 2 with only a small water level decline in the fault zone. Fault zones may also place highly fractured rock units in hydraulic connection with minable rocks. Faults can behave as recharge boundaries as shown in Fig. 14.

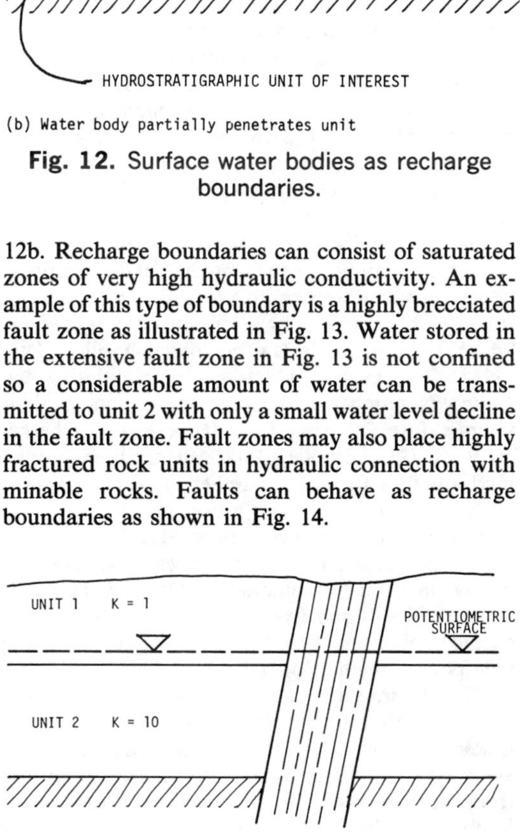

Fig. 13. Potential recharge boundary as brecciated fault zone.

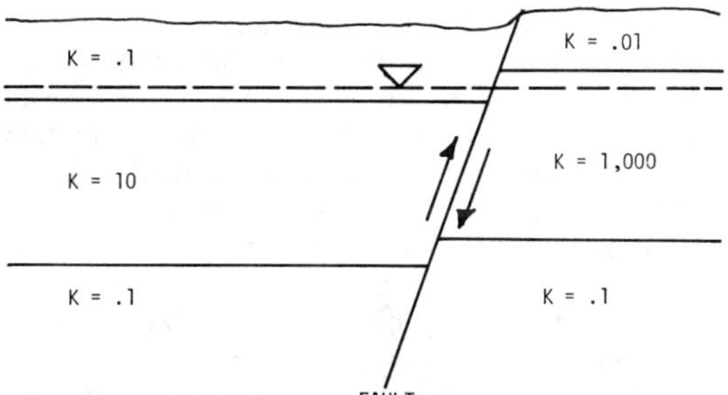

Fig. 14. Potential recharge boundary created by fault displacement.

Heterogeneity and Anisotropy

Frequently, the hydraulic conductivity of a hydrostratigraphic unit varies both with position in the unit (heterogeneity) and direction (anisotropy). Occasionally a unit will have a hydraulic conductivity that is independent of its position in the unit (homogeneity) but seldom is the hydraulic conductivity independent of direction (isotropy). In other words, the hydraulic conductivity parallel to the bedding and perpendicular to bedding may be the same throughout the unit but they may not be equivalent.

Heterogeneity and anisotropy in porous media such as valley fill deposits are caused by the processes of deposition. Depositional processes cause coarse material to accumulate near its source while the finer material is carried farther down gradient. These same transport processes sort the material. Predominantly coarse material is deposited with medium to fine grained material near the source while only the very fine material is carried great distances. A gradational decrease in the hydraulic conductivity parallel to bedding would be expected to occur down gradient. Layered heterogeneity occurs frequently in valley fill deposits. High hydraulic conductivity units tend to have much greater lateral continuity than vertical continuity. Anisotropic conditions result with higher horizontal hydraulic conductivity than vertical hydraulic continuity.

Cementation and consolidation reduce the porosity and hydraulic conductivity of consolidated sedimentary units. The grain size and size distribution are the primary factors controlling the magnitude and distribution of the hydraulic conductivity in a given sedimentary unit but the additional effect of cementation and consolidation can alter those properties. The porosity and hydraulic conductivity of an igneous or metamorphic unit can be altered similarly. Consolidation may not be equal over all areas of a given unit due to variations in overburden pressure or compressibility of the unit. These variations result in a heterogeneous distribution of hydraulic conductivity. Cementation can occur from a simple welding of grains due to heat and pressure but cementation generally results from the deposition of calcium carbonate, silica, or iron oxide around the grains and filling the pore spaces. The nonuniform action of these physical and chemical processes results in a heterogeneous and anisotropic distribution of the hydraulic properties in the unit of interest.

Secondary porosity frequently provides the only significant hydraulic conductivity in a consolidated unit (Lohman, 1972, p. 4). The primary porosity and hydraulic conductivity can be altered greatly by fracturing and faulting. A competent unit will fracture when stressed while an incompetent unit will yield plastically. The thickness of a unit is important since the competency of that unit increases as its thickness increases even though the compressive strength is the same (Billings, 1954, pp. 89–90). Some materials such as limestone may heal after fracturing or faulting, thus negating a potentially significant increase in hydraulic conductivity. Fractures and faults frequently provide conduits for the percolation and movement of water that can be slightly acidic. A common form is carbonic acid that will dissolve limestone; solution-widened channels increase both the storage capability and the hydraulic conductivity of the hydrostratigraphic unit.

Secondary mineralization can also provide a means of increasing the primary porosity. An example is the alteration of calcite to dolomite. The resulting decrease in the size of the crystal lattice is approximately 13% so there is an increase in the porosity and hydraulic conductivity of the unit (Freeze and Cherry, 1979, p. 154).

Zones where sudden, large inflows occur are of great significance in mining. These zones, sometimes referred to as pressurized zones, are usually created by a heterogeneous distribution of hydraulic conductivity that may be isotropic or anisotropic. A mine may encounter a zone where sudden, large inflows occur when progressing in a unit with a relatively low hydraulic conductivity. Often the mine has penetrated a zone with a relatively high hydraulic conductivity that may have nearly the same potentiometric level or one significantly higher than in the low hydraulic conductivity unit. Usually, the rapid increase in hydraulic conductivity is the real cause of large inflow and not "pressurized" areas although the interception of an anomalous confined (artesian) hydrostratigraphic unit can occur. Mine water inflow in such situations may initially be very high, on the order of several thousands of gallons per minute, but the inflow usually declines with time to small quantities as storage is depleted. The discharge characteristics are dependent on the hydraulic conductivity, coefficient of storage, fluid potential, and areal extent of the "pressure" zone. A unit with a small amount of water in storage may yield a large volume of water for only a brief period. Conversely, a high inflow rate may be maintained with time if there is a large volume of water in storage or if the zone is hydraulically connected to a source of recharge. "Pressurized zones" include fault zones, fault displaced units with high hydraulic conductivity, individual fractures, complex fracture systems, and discontinuities. Delineation of these features during premining investigations are essential to the prediction of mine water inflow.

Fractured Media

Ground water flow through fractured media requires special consideration. A system of discontinuous fractures may not increase the storage characteristics of the unit significantly, but even a single, extensive fracture can significantly alter the apparent hydraulic conductivity of a unit and result in significant mine water inflow. A fracture or fracture system can increase the primary hydraulic conductivity and create heterogenetic and/or anisotropic conditions. Such an occurrence might necessitate defining the fractured zone as a separate hydrostratigraphic unit independent of the non-fractured zone even though the rock may be mineralogically identical.

A principal concern when considering flow through fractured porous media is the point at which turbulent flow commences. Velocities in fractured media can easily attain turbulent flow, at which point Darcy's law no longer applies. Turbulent flow may be identified by calculating the Reynolds number that is the ratio of inertial to viscous forces during flow (Freeze and Cherry, 1979, p. 72). Turbulent flow can occur in porous media with a Reynolds number of 60 (Domenico, 1972, p. 148).

General porous media flow equations may be applied in many fracture flow cases by viewing the media on a macroscopic scale. On this scale, the fracture blocks and fractures represent the microscopic grains and flow paths associated with the true porous media flow (Parsons, 1966).

The hydraulic conductivity of fractured media has been estimated based on aperture size and joint spacings (Snow, 1969). Major difficulties arise when attempting to determine hydraulic conductivity from fracture geometry. The accurate location of fractures in the area of interest is very difficult as is the determination of aperture or fracture width on an acceptable scale.

The removal of fluid from a fractured porous media in which the unfractured rock has primary hydraulic conductivity initially derives flow from the fracture due to its much lower resistance to flow. This removal of fluid creates an imbalance in the fluid potential in the vicinity of the fracture with a greater fluid potential existing in the pores of the rock mass than in the fractures. Flow occurs from the pores to the fracture until the fluid potentials are equal. The initial reduction in the fluid potential of the fracture can cause the fracture aperture to decrease. A decrease in the hydraulic conductivity of the fracture results (Freeze and Cherry, 1979, p. 74). Long periods of time are required for the dewatered fracture system to equilibrate hydrodynamically with the pore space fluid potential distribution. It is for this reason that water content calculations for underground oil shale retorts cannot be based on water removed from fracture systems in oil shale hydrostratigraphic units.

Alteration of Flow System Variables

Surface and ground water flow systems can be altered significantly by underground mining. The formations overlying a mine can settle or collapse, depending upon the properties of the porous media and the mining schemes used at the site. Surface water basin divides can be altered and new subbasins formed by this settlement or collapse. Virtually direct conduits for flow have formed between surface water systems and mine workings where caving has extended to land surface.

The ground water system is affected directly by mining if the mining occurs in saturated formations (Fig. 15). Each shaft and tunnel creates a path of essentially infinite hydraulic conductivity when compared with the surrounding material. The overall hydraulic conductivity when viewed on a macroscopic scale is increased greatly. Existing ground water flow patterns are altered greatly by these paths of high hydraulic conductivity. The storage properties are altered since the drifts and shafts have a 100% porosity. Large volumes of water can be stored in flooded shafts and drifts. These storage areas can be hazardous during the reopening of old mines that have not been bulkheaded properly and mapped adequately.

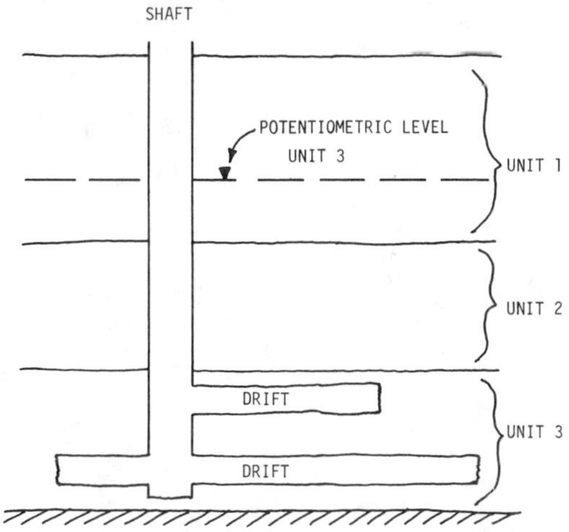

Fig. 15. Impact of mining on hydraulic conductivity of a hydrostratigraphic unit.

3

Data Requirements

GENERAL STATEMENT

The primary objectives of any field data gathering effort should be to (1) identify and gather the data necessary for the project and (2) obtain the data in a state-of-the-art manner. All too often the initial field data are collected both areally and temporally in an illogical manner without the guidance of a conceptual model of the ground water flow systems involved or even a review of existing geologic literature on the area of interest. The initial data collected frequently are of limited value while necessary basic reconnaissance information is missing.

Initial field data should be collected with the intent of developing a hydrologic overview of the potential mine site and surrounding area. Obviously, one of the initial objectives is to define the area requiring a hydrologic investigation. The data requirements should be identified by the time frame in which collection should be made and by the corresponding increase in sophistication of the data requirements with development and operation of the mine. The data requirements are summarized in Table 1.

INITIAL LEVEL SITE INVESTIGATION

Area Determination

The initial task of any hydrogeologic investigation is to determine the boundaries of the area requiring study. Obviously, the site of the proposed mine is included in the study area. The areal extent beyond the site may be determined from an evaluation of existing geologic and topographic maps. Those formations that overlie the ore body, the formations containing the ore body, and the formation(s) that lies immediately beneath the ore body are of direct concern for proper site reconnaissance. Additional formations below the ore body may require study depending upon their thickness, hydraulic conductivity, and degree of interconnection with the mine workings. This initial viewpoint identifies hydrostratigraphic units based strictly on geologic concepts such as mineralogy and structure. Formation outcrops, synclines, anticlines, faults, and fracture and joint patterns are used to delineate the area of the site reconnaissance.

The simplistic hydrogeologic environment (illustrated in Fig. 3, chapter 2) requires that field data be collected via test wells and/or geophysical techniques. This approach is necessitated by the lack of surface features such as formation outcrops, streams, and springs. Fig. 5 (chapter 2) illustrates a slightly more complex hydrogeologic regime. The potential mine sites at locations A, B, C, D, and E each intercept a different ground water flow system or combination of flow systems. Therefore, each mine location requires that a different area and size of area be investigated.

A more complex geologic setting as illustrated in Figs. 6 and 7 (chapter 2) may be approached differently. The area included for the site reconnaissance should encompass sufficient surrounding area to include the outcrops of those formations suspected of being influenced by the future mine. Even adjacent areas not suspected of being influenced may be investigated if the formations of interest crop out in those areas. Such an extension of the area of investigation would provide a greater regional understanding of the hydrogeologic properties of the formations (hydrostratigraphic units) of interest.

Geologic Investigation

The initial step before conducting the site reconnaissance is to review all existing literature on the geology of the area. Existing information should be augmented with new exploration data on the dip, strike, thickness, and lateral extent of the formations in the area. Exploration hole logs should be reviewed for indications of lost circulation, rubble zones, and water producing zones. Existing aerial photos such as those available from the US Department of the Interior, EROS Data Center,

Table 1. Data Requirements Summary

Initial level site investigation	Second level site investigation	Data collection after commencement of mining
Area determination Compare geologic and topographic maps with mine plan	Spring and stream data 1) Collect time-dependent data on spring discharge 2) Collect time dependent data on gaining reaches of streams	Geologic data Update hydrostratigraphic unit identification and location
Geologic investigation 1) Review existing literature 2) Complement with exploration and new data	*Potentiometric level data Collect time-dependent water level data from test wells (appropriately located with respect to hydrostratigraphy)	Surface water data 1) Collect precipitation data 2) Collect spring flow data 3) Collect stream flow data
Hydrologic investigation 1) Determine areal distribution of precipitation (temporal distribution may also be required) 2) Locate springs and measure discharge, temperature, and mineralization (water quality) 3) Measure temperature, mineralization (water quality), and rate of flow of streams at formation contacts 4) Locate existing wells and available data on same	*Hydraulic conductivity data 1) Determine hydraulic conductivity, by hydrostratigraphic unit, parallel and perpendicular to bedding, preferably via field testing	Potentiometric level data Collect time-dependent potentiometric data by hydrostratigraphic unit (most frequent measurements near mine)
	*Storage characteristics 1) Determine specific yield (unconfined storage coefficient) of those hydrostratigraphic units that will be completely or partially dewatered 2) Determine storage coefficient of confined hydrostratigraphic units that can be influenced by mine inflow; field tests preferred	Mine inflow data 1) Record mine inflow 2) Note hydrostratigraphic unit correlation to sudden, large inflows 3) Monitor rate and duration of inflows

* Data mandatory for mine inflow predictions.

can provide information for surface geologic mapping and interpretation of geologic features such as linears.

Formations should be identified and mapped in the field where existing data are lacking. The dip and strike of these formations facilitate the interpretation of geologic structure that can play a significant role in controlling ground water flow systems. Well logs from existing wells in the area occasionally can provide useful information on subsurface data and aid in the interpretation of the surface data.

Hydrologic Investigation

Ground water flow systems can be formulated if the recharge and discharge areas can be identified. Recharge is from precipitation; therefore, the areal distribution of precipitation must be known in at least a qualitative sense. This distribution may need to be augmented by a qualitative knowledge of the temporal distribution of the precipitation, since local ground water flow systems may exist for only a short period of time after a precipitation-related recharge event occurs. Correlating the temporal distribution of precipitation with noted discharge areas aids in the formulation of conceptual models of ground water flow systems. This procedure is essential to predicting mine water inflow because under water table conditions where recharge occurs, the majority of mine water inflow may occur over only a short period of time in response to recharge events.

Springs are one form of ground water discharge that should be mapped during the site reconnaissance. Their location should be recorded with respect to geology, using both formation and structure. Spring discharge should be measured for rate of flow as well as temperature and quality of the water. Both the temperature and the quality of the water will reflect the type of ground water flow system associated with the spring.

Stream flow measurements can be a useful tool for predicting which formations can support ground water flow systems. Ideally, a section of stream that flows across the complete sequence of formations (hydrostratigraphic units) that potentially can affect mine inflow is monitored (Fig. 16). The ground water contribution to stream flow or the stream flow addition to the ground water system is easily determined by comparing stream flow measurement data at each formation or hydrostratigraphic unit contact. Gaining reaches of the stream indicate that the stream channel is a discharge site for the formation exposed within that reach. Losing reaches indicate that the formation exposed in that reach is being recharged with stream flow. Both cases indicate that the formation has "some" hydraulic conductivity. A lack of change in stream flow across a formation may be an indication that the formation has a very low hydraulic conductivity

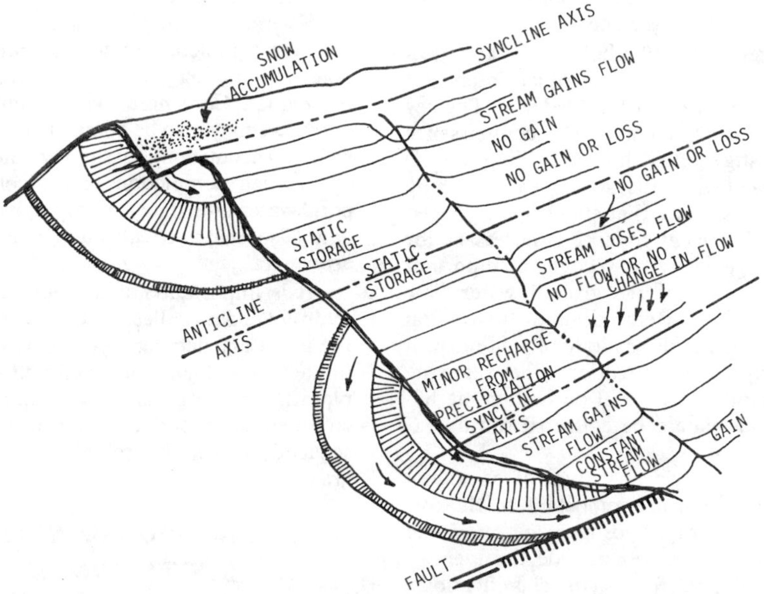

Fig. 16. Orthographic view of stream gain-loss characteristics in the southeast Idaho phosphate district (Ralston, et al., 1980). Legend: hatched area, low hydraulic conductivity; plain areas, high hydraulic conductivity; dot-dash lines, syncline or anticline axis.

in comparison to the previously mentioned situations. Such a categorization requires an evaluation of the geology in the area with respect to hydraulic concepts. Ralston (et al., 1980, pp. 38–41) describes the evaluation process of stream gain-loss data in greater detail.

Increases in stream flow may also be detected by measuring the temperature and mineralization (total dissolved solids content) of the stream water at specific locations. These procedures assume that there is a significant difference in the quality of the surface water and the ground water and that any ground water contribution is significant enough to be monitored by these means. This procedure is particularly useful in locating structurally controlled hydrostratigraphic units within a single formation.

Hydrogeologic data collection for the site reconnaissance level should consist of locating existing wells and evaluating available well logs. Significant quantities of data on ground water levels usually do not exist at this stage. Often existing wells were constructed so that they intercept several formations. Data from such wells do not accurately represent a single formation or hydrostratigraphic unit. Frequently, the wells are open to aquifers with differing potentiometric levels; the resulting measured water level is thus a composite of the potentials. The log for the well should be evaluated to delineate hydrostratigraphic units if possible.

Hydrogeologic data collection should not be overlooked during the exploratory phase of the geologic investigation at a mine site.

In some cases exploration holes drilled for geologic purposes also can provide other important hydrologic data with very little added cost. The key factors in successfully combining reconnaissance hydrologic investigation with core drilling are (1) an exploration geologist with ground water training and (2) an outline of data that should be collected during drilling. Sufficient training is necessary for the exploration geologist to recognize important general characteristics of the ground water flow system at the mine site. Recognition of the general characteristics of the hydrogeology of a site is necessary for effective data collection. The following paragraphs constitute a general outline of the hydrologic data that should be collected during exploration drilling.

Many of the geologic objectives of an exploration drilling program are important to the formulation of a conceptual hydrogeologic model of a mine site. These objectives include the identification of rock types, formation classifications, strike and dip, degree of induration, and degree and character of fracturing. Secondary hydraulic conductivity created by fractures and joints is a principal controlling factor for water inflow to many mines. Hydrogeologic data may be obtained both from examination of core and from monitoring during drilling. Identification of fractures and joints and their spacing from cores provides important input to classification of the rock with respect to hydraulic properties. The chemical analysis of the cores can yield important data with respect to mine water chemistry. For example, ratios of pyrite content to calcite content can be used as an indicator of potential acid-producing conditions and buffering conditions.

The drilling of a core hole provides an approximate in situ test of the rock penetrated by the hole. Fluctuations in the level and consistency of drilling mud provide important clues with respect to hydraulic properties of rocks penetrated. For example, lost circulation zones often indicate zones with high hydraulic conductivity and perhaps lower hydraulic head. Notes should be taken that identify specific lost circulation zones and their characteristics. The thinning of drilling mud can indicate a zone with both high hydraulic conductivity and high hydraulic head.

Optimum hydrologic data collection occurs when drilling muds have viscosities and densities approaching fresh water. Such drilling conditions facilitate measurement of hydraulic head with depth during drilling. The depth to water should at least be measured at the start of each day's drilling. Certainly flowing conditions should be noted because they have implications with respect to mine water inflows.

Hydrologic data collection does not have to end with the completion of core drilling. Selected core holes may be logged using a range of geophysical techniques. Core holes with significant water occurrences may also be completed as small diameter wells (piezometers), assuming the holes remain open. Such completions allow collection of temporal water level data (hydrographs). Water quality data also may be collected from completed piezometers.

It is important not to confuse reconnaissance hydrologic data collection during core drilling with the necessary detailed hydrologic study. However, limited hydrologic data collected early in the exploration stage in mine development can save a considerable amount of money in designing the subsequent detailed hydrologic data collection program.

INITIAL GROUND WATER FLOW SYSTEM EVALUATION

Hydrostratigraphic Unit Identification

Data from the initial level site investigation are used for the first identification of ground water flow systems. Hydrostratigraphic units are defined as the first step in this process since the site reconnaissance data were collected based only on geologic concepts. It should be noted again that hydrostratigraphic units are defined by hydrogeologic properties that may not coincide and may in fact cross the more common geologic definition of units or formations. Borehole geophysical logs and cores from exploration drill holes may assist in the delineation of hydrostratigraphic units. These data may be particularly helpful if the exploration crew has been alerted to the needs of the mine hydrologist and mine planner. Exploration personnel with training in mine hydrology are especially useful in this regard.

The geologic setting of a spring should be reviewed with regard to stratigraphy and geologic features such as joint systems, fractures, and faults. Correlation with these features should be noted and further evaluated, based on the magnitude of the spring discharge. Springs should be grouped by the formation from which they emanate, whether or not the spring is associated with a geologic feature such as a fault, and by the magnitude of the discharge.

Stream gain-loss data should be gathered and similarly evaluated. Those sections of the stream that gain or lose flow can be correlated in a similar manner as the springs. Sections that do not have a change in flow must be evaluated, based on potential recharge or discharge sites and hydraulic continuity. Sections of the stream should be grouped by their gain, loss, or no change in flow characteristics, formation, geologic features, and by the magnitude of change in flow.

The grouped data for the springs and streams should be compared for common features. Geologic formation and stratigraphic unit correlation is the most obvious comparison to make. Those springs influenced by features such as fault zones should be evaluated separately. The number of springs and their discharges should be compared by formation or unit and separately by other geologic features. A pattern should develop from evaluating the data in this manner.

A structural feature such as a fault zone may have distinctly different hydraulic properties at different sites, depending upon the formation it crosses. Defining such a feature as an individual hydrostratigraphic unit must be based upon consideration of the location of springs and their discharge and the gain-loss characteristics of streams crossing the fault zone.

A preliminary definition of hydrostratigraphic units should be possible at this time, based on spring discharges and stream gain-loss characteristics. Formation boundaries may or may not reflect hydrogeologic properties. One hydrostratigraphic unit may be identified by a lack of springs or stream gain or loss. Another hydrostratigraphic unit may support numerous small springs and moderate gains in stream flow. A third unit might be designated because there are large losses of stream flow across some exposures and large springs at others. Such a situation could lead to a classification of the units in relative approximate terms such as low, medium, and high hydraulic conductivities.

Conceptual Model of Ground Water Flow Systems

A conceptual model of the ground water flow systems near the future mine site may now be developed, utilizing the data from the techniques discussed previously. Recharge to a ground water flow system will occur primarily where the hydraulic conductivity of the exposed or near surface hydrostratigraphic unit(s) is medium to high (Fig. 4, chapter 2) and there is a source of water such as a stream. Those units with a low hydraulic conductivity may store large amounts of water; however, these units cannot support a significant ground water flow system.

The topography of the potential recharge sites and the geologic-structural controls on the position and attitude of the hydrostratigraphic units should be noted. The hydraulic continuity of potential ground water flow paths must be evaluated from the recharge site(s) to a potential discharge site(s). The elevation of the flow path in a given hydrostratigraphic unit and potential discharge site(s) must be lower than the elevation of the static water level in that unit at the recharge site (Fig. 17). Ground water flow in Fig. 17 can proceed from ridge B, where recharge occurs, across ridge A but not across ridge C, due to the elevation of unit 2 at ridge A and ridge C.

Potential discharge sites should be compared to discharge sites noted in the field such as springs and gaining reaches of streams. The relative water quality of the springs should be compared to the length of the conceptual ground water flow paths. The longest paths should correspond to the highest total dissolved solids content and the highest temperatures. Water temperatures near the mean annual air temperature and low total dissolved solids contents should correspond to the shortest conceptual flow paths.

Ground water level data from existing wells can be grouped into two categories, assuming that the data can be assigned to specific hydrostratigraphic units. Water level data for the same unit over a broad area of the site can be used to determine the lateral direction of ground water movement in that unit. Water level data in a small area for several units can be used to describe vertical hydraulic head and thus vertical movement of water between units. Both types of ground water data aid in the delin-

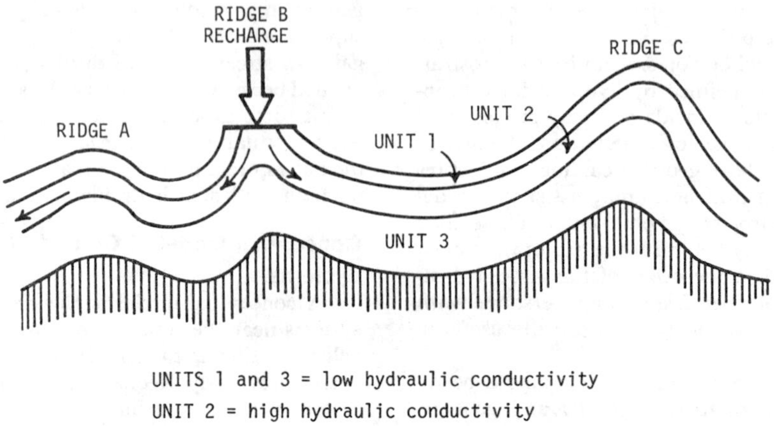

Fig. 17. Elevation control of ground water flow paths. Units 1 and 3 have low hydraulic conductivity. Unit 2 has high hydraulic conductivity.

eation of ground water flow paths and areas of recharge and discharge.

The evaluation of all of the foregoing factors facilitates the formulation of a conceptual ground water flow system involving all the units of interest at the mine site. Anomalies will doubtless exist at this stage where conceptual flow systems do not coincide with actual recharge and discharge sites and their apparent flow systems. These anomalies should be noted and the geology in that area reevaluated. Additional spring and stream data should be collected in the anomalous area if not covered adequately during the initial level site investigation. The precipitation pattern also may require a more detailed analysis. The conceptual flow systems should be redefined based on these additional and more detailed data if warranted.

COMBINED MINE PLAN AND CONCEPTUAL MODEL

The proposed mine plan should now be combined with the conceptual model just formulated. In reality the two should be developed simultaneously because the mine plan cannot be developed without the conceptual model. It should change as the conceptual model changes with the acquisition of more data. The mine plan may be imposed in stages such as placing the first shaft upon the conceptual model followed sequentially by additional shafts and drifts. Each addition of the mine plan should be viewed with regard to the hydrostratigraphic units and the resulting ground water flow systems that the mine will intercept. Those units that are sealed off by grouting or other means can be eliminated from consideration. The relationship of the mine to the ground water flow systems should be noted as illustrated by the simple system shown in Fig. 5, Chapter 2. Alternative mine plans can result in a wide range of impacts on the ground water flow systems. The potential effects of mining on surface water systems can be evaluated based on those ground water flow systems that will be affected by the mine.

A mine that will affect local ground water flow systems will only affect those surface water systems associated with the local ground water systems. Mines affecting intermediate and regional ground water flow systems will affect the associated surface water systems that will occur at greater distances from the site. The increasing distance from the site will cause a concurrent increase in the time required for a mining-induced impact to occur on the surface water system. The imposition of a mine in a natural hydrogeologic environment may alter both intercepted and adjacent ground water flow systems. Adjacent flow systems should be included when developing the second level site investigation.

SECOND LEVEL SITE INVESTIGATION

This phase of the site investigation includes subsurface information that can be obtained only by drilling test wells and installing water level monitoring devices. Some sites, as noted previously, will require that the initial site investigation begin with this subsurface acquisition of data without benefit of the surface indicators of hydrogeologic properties. Additional hydrologic data may be obtained by reviewing the geologic drilling logs in the area from both water wells and geologic exploration holes. Ideally, hydrogeologic data were collected during exploration drilling. Lost circulation, water producing zones, and rubble zones are examples of the type of indicators identified when reviewing this type of data.

Field work for this level of site investigation should be preceded by an evaluation of the conceptual model. Additional data collection should be concentrated in the areas where potential mine impacts can be foreseen. Data collection sites peripheral to this area and environmentally sensitive areas should be monitored continually for background information.

Spring and Stream Data

Spring discharge data should be collected on a time-dependent basis during this level of site investigation. The data collected should be evaluated for the recession flow characteristics of the springs. Suggested references include Domenico (1972) and Linsley et al. (1958). Gaining reaches of streams should be monitored similarly so that variable ground water contribution to the stream can be recorded. Both forms of data will be used for computer model calibration.

Potentiometric Level Data

Time-dependent water level data from test wells should be collected at this stage. The time intervals between field measurements will depend on several factors, including the depth of the flow system being monitored, distance from the recharge area, and the time since recharge occurred. Initially, data should be collected frequently so that trends can be determined accurately. The time intervals can be lengthened as confidence in the trends increases or the importance for procuring highly accurate data declines. Monthly monitoring of wells should prove adequate for the initial period of measurement of ground water levels. Subsequent time intervals may be extended to quarterly or even semiannually. The semiannual readings should correspond to the highest and lowest anticipated ground water levels. A few key wells should be fitted with continuous water level recording devices so that a continuous record of water level fluctuations is obtained.

The vertical distribution of hydraulic head between hydrostratigraphic units must be measured, especially in recharge and discharge areas. Knowledge of the distribution of hydraulic head at these locations aids in the determination of vertical rates of movement between units and in the calibration of the mathematical model. Measurement of the head or potential distribution in the mine area is necessary both for defining premining vertical rates of ground water movement and as a basis against which mining impacts can be judged.

Hydraulic Conductivity Data

The distribution of hydraulic conductivity in each hydrostratigraphic unit must be determined for two different orientations. Hydraulic conductivity values must be determined both parallel and perpendicular to stratification, especially in the recharge and discharge areas. Computer model water levels are particularly sensitive to the ratio of hydraulic conductivity parallel and perpendicular to stratification in these areas. The hydraulic conductivity must also be defined in the mine area for both orientations so that mine inflow predictions can be made. Hydrostratigraphic units may require redefinition after obtaining hydraulic conductivity values.

Hydraulic conductivity values can be determined from a number of procedures, including laboratory and field techniques. Core samples frequently are obtained during the site investigation for ore samples. These cores can be evaluated visually for their apparent hydraulic conductivity and suitability for laboratory testing. A few selected sections of core could be tested for hydraulic conductivity. It should be noted that these sections of core are only representative of a very small volume of a hydrostratigraphic unit and seldom give any indication of the effects of secondary porosity such as fractures.

In situ field testing with single wells can provide a reasonably inexpensive procedure for determining the hydraulic conductivity of a hydrostratigraphic unit. Single well tests are more representative of actual field conditions than core tests conducted in a laboratory. The radial distance from the well, and thus volume of material tested, that is actually influenced by the test depends on the duration and type of test used. These tests are more likely to reflect the effects of secondary porosity features such as fractures.

Testing with a pumped well and several observation wells permits the determination of a number of parameters, including vertical and horizontal hydraulic conductivity. The area influenced and monitored with this type of test is much greater than with the techniques previously described. The cost is also greater, especially as the depth of the units to be tested increases. Freeze and Cherry (1979) provide additional information on conducting such tests.

Storage Characteristics

The storage characteristic of the uppermost hydrostratigraphic unit is defined by its specific yield. Underlying units will usually be defined by the specific yield and the storage coefficient controlled by the compressibility of the aquifer and the fluid. Both values for storage will probably be required for the underlying units so that mine inflow predictions can be made.

Specific yield can be determined in a laboratory from undisturbed samples of the hydrostratigraphic unit. The storage coefficient can be calculated from a laboratory determination of the variables in Eq.

2 (page 8) but a field determination of these storage characteristics is preferable where possible. The specific yield for the uppermost hydrostratigraphic unit can be calculated from long duration pump tests conducted with a pumping well and at least one observation well. The specific yield of underlying units also can be determined from pump tests if the units form an unconfined ground water system. A value for the storage coefficient of the underlying units can be estimated from some types of single well tests; multiple well tests can provide the necessary data for calculating a more accurate and representative value for units with a confined ground water system. Borehole geophysical techniques also provide a means for estimating specific yield (Keys and McCary, 1976).

Evaluate Conceptual Model

Hydrostratigraphic units may require redefinition based on the specific hydrogeologic properties now described for each unit. Specific values for the hydraulic conductivity and storage characteristics probably will form the basis for the definition of new hydrostratigraphic units. Even a slight change in their definition can alter significantly the conceptual model of ground water flow systems. The conceptual model should be redefined as new data become available.

FIRST LEVEL APPLICATION OF ANALYTICAL AND NUMERICAL TECHNIQUES

Specific data on the hydrogeologic properties of the mapped hydrostratigraphic units now can be incorporated into the techniques available for predicting mine inflow. The analytical techniques (manually calculated procedures) can be employed or the data can be utilized in a numerical technique (computer model) or a combination of both procedures can be implemented. Both techniques are described in a subsequent chapter. The remainder of this section and subsequent sections of this chapter deal with the application of the computer model for predicting mine inflow.

Values for hydraulic conductivity parallel and perpendicular to stratification in hydrologic zones of influence of the proposed mines should be incorporated into the computer model for each hydrostratigraphic unit identified according to the procedures described previously. The storage characteristics for each unit should be incorporated also. Premining conditions only are modeled at this point. The model should first be run for a zero storage constant head situation depicting only areas of natural recharge and discharge and the head distribution between these areas. This computer simulation should be compared with the areas of known recharge and discharge and points of measured hydraulic potential. Those areas where the computer model deviates from the actual field situation should be noted. Alternative causes and the degree or extent of the deviations should be evaluated. The model should be capable of allowing subsequent adjustments in the potentiometric surface so that the model can be applied to different stages of mine dewatering at the second and third levels of analysis.

The hydrogeologic parameters inserted in the model are point values. Adjustments of the values used in the model between these points may adequately calibrate the model to comply with the known field situation. The input values of hydraulic conductivity are adjusted at this stage. Additional field data may be required to model more accurately the ground water flow systems around the mine site. The sequence described in this section for the steady-state model may have to be repeated several times if the hydrogeology is complex.

SECOND LEVEL NUMERICAL TECHNIQUE

The computer model at this point has been calibrated adequately to reflect the areas of known recharge and discharge and the fluid potential distribution. The second level model is run to simulate temporal changes in ground water levels and discharges under premining conditions. The fluctuations of the water level in response to recharge are largely a function of the storage properties of the units being monitored. Adjustments are made primarily in this parameter to allow the model to simulate the measured temporal changes noted for the ground water levels. Portions of the model simulation still may not reflect field conditions adequately. Potential causes for these variations should be evaluated and additional input data obtained for the model if warranted by the evaluation. This is the final procedure for modeling the premining ground water flow systems.

THIRD LEVEL NUMERICAL TECHNIQUE

The mine plan is now imposed on the computer model. The mine plan should be considered in stages so that the inflow can be predicted at critical points during the progress of the mine. Such critical points might be defined as evenly spaced periods in the mine development in a homogeneous unit or at times when mining proceeds to hydrostratigraphic contacts and/or boundaries in heterogeneous units. The point in time when the mine will enter a unit with a higher hydraulic conductivity is an important period for predicting sudden high rates of inflow. This critical point could be described as a "pressurized zone." Inflow predictions are obviously needed for mine planning.

Inflow of water into a mine and subsequent pumpage from the mine reduces the ground water levels in the area. The model should also be run for a prediction of water level declines in the various hydrostratigraphic units influenced by the mine. These declines can indicate potential areas where natural ground water discharge will be reduced or eliminated. The model should be run at discrete times selected for the complete development of the mine. Long-term impacts beyond mine abandonment can also be evaluated with the model.

DATA COLLECTION AFTER COMMENCEMENT OF MINING

Additional data should be collected as the underground excavation commences. The hydrogeologic framework should be updated with any new pertinent information dealing with the hydrostratigraphic units. The location of the hydrostratigraphic units should be noted as they are encountered. All previously unknown geologic structures should be incorporated. The model should be updated where the new data could influence significantly the predictive capabilities of the model.

Surface water data should be collected where there is a potential for interconnection with the ground water system. The temporal distribution of precipitation should be recorded routinely, especially for correlation to low spring discharges and stream baseflow. Spring discharges also should be monitored to facilitate the determination of mine-related impacts. Stream discharges should be monitored similarly, especially along gaining reaches of the stream.

Mine inflow should be monitored. The total inflow should be recorded as mining progresses. Detailed data concerning large, sudden inflows should be noted as to their source (hydrostratigraphic unit) and rate of inflow with time; these large inflows will almost always decrease with time. Ground water levels should be monitored with the most frequent readings being taken nearest the mine. Appropriate time intervals will depend upon the rate of change of the ground water levels. A rapid change in water levels indicates the need for frequent measurements and vice versa.

4

Techniques of Data Collection

GENERAL STATEMENT

Data frequently can be collected by several different methods. The method required for the necessary precision for inflow prediction to mines should be selected and other methods avoided. Precision monitoring or measuring techniques frequently require more time, expertise, and expense in their implementation that may not be justifiable. The temporal distribution of the data needs also may dictate the method used for data collection. Continuous recording devices may be required for time sensitive data whereas point measurements of other parameters may be adequate. Data needs will vary as the hydrologic investigation proceeds. Initial data needs may require only a qualitative collection of data such as the areal distribution of precipitation. Subsequent data needs may require the measurement of ground water levels to the nearest hundredth of a foot.

Data requirements can be difficult to fulfill for ground water systems. This is particularly true for anisotropic, heterogeneous hydrostratigraphic units since the hydrogeologic parameters vary both vertically and areally. Techniques exist for determining the storage characteristics of both unconfined and confined hydrostratigraphic units but these procedures require the installation of multiple wells in the same unit which may be impractical. Specific tests are discussed subsequently that permit the determination of vertical hydraulic conductivity.

Data collected from field procedures or tests are generally preferred to laboratory derived data. This is especially true for hydrogeologic parameters such as hydraulic conductivity.

SURFACE WATER SYSTEMS

Precipitation

Precipitation data can be collected from sources that maintain a historical record. The US Department of Commerce, National Oceanic and Atmospheric Administration (NOAA), maintains weather observation stations across the country. Such a station frequently can be found near a proposed mine site but the type of weather data collected varies from station to station. Fortunately, precipitation data are collected ubiquitously. The data collected by NOAA can be purchased for the area of interest in the form of monthly and annual tabulations. Private weather stations constitute a second source of precipitation data.

Existing data usually are analyzed by one of three methods. The station average method merely sums the depth of precipitation at each station and divides the sum by the number of stations as follows:

$$P_{avg} = \frac{\sum_{i=1}^{N} P_i}{N} \tag{3}$$

where P_{avg} is the average precipitation (L), P_i is the depth of precipitation at station i (L), and N is the total number of stations. This method is applicable where there are a relatively large number of stations uniformly distributed across the area (Hjelnfelt and Cassidy, 1975, pp. 71–72).

The second method, called the Thiessen polygon method, weights the precipitation at each station in direct proportion to the area it represents without consideration of topography. The stations are plotted on a map and perpendicular bisectors are drawn on the lines connecting the stations. The polygons formed by these bisectors constitute the area that the precipitation for that station represents. The overall precipitation can be calculated with the following equation:

$$P_{avg} = \frac{\sum_{i=1}^{N} P_i A_i}{\sum_{i=1}^{N} A_i}\ , \tag{4}$$

where A_i is the area assigned to station i (L^2) (Hjelnfelt and Cassidy, 1975, pp. 71–72). All other variables are as described in Eq. 3.

The isohyetal method is the most accurate method for averaging precipitation over an area (Linsley, et al., 1958, p. 82). This method is the only technique that allows consideration of orographic or other topographic effects in the computation of an overall precipitation value. Contours of equal precipitation are drawn on the map using the data from each station. The area within each set of contours is measured and multiplied by the average precipitation for that area. Each area is summed and the total averaged by the following equation:

$$P_{avg} = \frac{\sum_{j=1}^{m} \left[(P_j + P_{j+1}) \left(\frac{A_j}{2} \right) \right]}{\sum_{j=1}^{m} A_j} \quad (5)$$

where P_j is the precipitation at contour j (L), A_j is the area enclosed between contours j and $j+1$ (L^2), and m is the total number of contours (Hjelnfelt and Cassidy, 1975, p. 73).

Precipitation can be measured by any of several different methods. The simplest and most common is the standard 8-in. (20.3 cm) precipitation gage. The 8-in. (20.3 cm) collector on this model concentrates the flow in an inner cylinder where the depth of precipitation can be measured manually. Recording gages also are used in three common forms. The tipping-bucket gage consists of a collector and a two-compartment bucket, each with a capacity of 0.01 in. (0.025 cm). As one bucket fills and pivots to tip the water into a reservoir, the second bucket pivots beneath the collector. Each tipping motion activates an electrical circuit, causing a pen to mark on a revolving drum. The weighing type recording gage is available as a second alternative. The precipitation falls into a bucket set on a platform of a spring or level balance. The increasing weight due to precipitation is recorded on a chart. Float recording gages are the third major type used. The equipment can vary in that the float may rest in the receiver that receives the precipitation or the float may rest in an oil or mercury bath. The receiver displaces the alternate fluid that is measured in the latter case. Increasing precipitation is recorded on a chart although alternate forms of recording the data can be used (Linsley, et al., 1958, pp. 66–68).

Precipitation can be measured by storage gages. This method is well suited for use in remote areas or where only the semiannual or annual accumulation of precipitation is of interest. These gages are precharged with an antifreeze solution to prevent freezing and oil to prevent evaporation losses. Measurements are made with a tape or by pumping out the contents for weighing.

Snow measurements can be made intermittently or continuously. In simplest form measurements can be made with a yard stick or equivalent measuring stick where the depth of accumulation is low. Graduated markers called snow stakes also can be used or aerial snow-depth markers for visual observation from low flying aircraft. Aerial photos can be interpreted for the areal extent of the snow cover, but the depth of the snow pack cannot be determined adequately from this method. The most important consideration in snow studies is not the depth of the snow pack but the water equivalent of the snow. Both pieces of data are needed to determine the amount of water held in the snow. Sampling for the water equivalent can be accomplished with a snow tube. This hollow tube is driven vertically through the snow pack to its base and the depth noted. The sections of the now snow-filled tube are withdrawn and weighed and the water equivalent calculated. Sample points are established along a snow course for additional runs throughout the season (Linsley, et al., 1958, pp. 92–94).

A continuous record of the water equivalent of snow can be obtained with a pressure pillow. The increasing weight of the snow on the pillow causes an increase in the fluid pressure within the pillow that can be measured easily. Pressure pillows are not subject to wind-induced errors as are other precipitation gages but they are subject to bridging. An ice crust can form over the pillow and cause false readings. Pillows are also vulnerable to damage by humans and animals (Linsley, et al., 1958, pp. 94–95).

Nuclear devices also are used to measure the water equivalent of snow. One technology suspends the radiation source in the air above a detector at ground level. The moisture equivalence is determined from the attenuation of a beam of neutron radiation by the snow. An alternative method utilizes twin vertical tubes approximately 2 ft (0.6 m) apart. A gamma photon source is lowered down one tube while a detector is lowered synchronously down the other tube. This type of gage is called a profiling radioactive snow gage or twin-probe snow-density gage (Linsley, et al., 1958, pp. 94–95).

Stream Flow or Spring Discharge

Surface water flow measurements can be divided into two categories; these are periodic measurements and continuous measurements. Periodic measurements can be made with a variety of portable flumes. A flume consists of a specially shaped open channel flow section that can be installed in a channel for measuring the rate of flow. The more common variety of portable flumes includes the V-notch, trapezoidal, and Parshall flumes. All fit in the general category of a Venturi flume in that flow through the flume is constricted in a controlled manner by tapered sidewalls or by varied elevation of the flume floor, or both (US Bureau of Reclamation, 1975, p. 85). These flumes can be installed in a semi-permanent to permanent fashion or they can be installed for each flow measurement. The periodicity of the measurements and complexity of the setup dictate the method of installation.

Flumes can be used to measure flow rates from less than 1 gpm (0.06 L/s) to several thousands of gallons per minute, depending upon the type and size of the flume selected. Higher rates of flow preclude the practical installation of a large flume for single flow measurements. Channel shape, slope, and bank materials also influence the practicality of installing and utilizing flumes effectively.

Flow technologies are based on head measurements within the flume. Some flumes require both an upstream and downstream head measurement on either side of the flow constriction. These head measurements are compared with a rating chart for the specific flume in use; flow rates are read directly from the chart. Excessive submergence of the downstream end of the flume can necessitate an adjustment to the calibrated flow rate.

A semipermanent to permanent installation can be made with a weir. Weirs can be used most effectively when there is a fall of 0.5 ft (0.15 m) or more available in the stream or spring discharge channel (US Bureau of Reclamation, 1975, p. 7). A weir is an overflow structure built across an open channel with specific dimensions and shape. Flow rates are determined from the height or head of water between the crest of the weir and the free surface of the pool upstream of the weir, the weir type and size, and flow charts or formulas.

Weirs generally are constructed in one of three basic shapes referred to as rectangular, Cipolleti or trapezoidal, and triangular or 90° V-notch. An additional type combines two types of weirs such as the 90° V-notch weir and the rectangular weir for special flow measurement requirements. The 90° V-notch weir measures the low flow rates while the rectangular portion is suitable for high flow rates. A small range of discontinuity occurs as the flow rate exceeds the 90° V-notch section and begins to flow over the broad crest of the rectangular section of the weir. Weirs are further classified by the type of weir notch used. A sharp edged weir plate can be used to prevent the flowing water from contacting the edges of the bulkhead. This type is referred to as a sharp crested weir. A broad crested weir is used where the wall or bulkhead is too thick for the flowing water to "spring past" the bulkhead. Discharge coefficients for the broad crested weir are usually obtained by calibrating the weir in place with some other method of measurement.

Numerous tables exist for determining the flow rate over a weir. Entrance velocity and submergence correction formulas and flow rate formulas exist for numerous types of weirs that are not covered by published tables. Installation guidelines are available from many sources (US Bureau of Reclamation, 1975).

Stream flow or spring discharge can be measured by manual metering techniques. The most common current meters in use include the Price-type meter and the Pygmy meter. Both meters consist of a series of cups mounted on a vertical axis. A make-break electrical contact is located on the top of the vertical axis. The Price-type meter frequently has two separate contacts; one contact counts every fifth revolution and one counts each revolution. The Pygmy-type meter has a single revolution contact (US Bureau of Reclamation, 1975, p. 118). A digital counter or battery powered headphone is connected to the electrical contact on the meter for counting revolutions within a known time interval. These data can be converted to fluid velocity with the calibration table for that meter. Fluid velocities are measured at prescribed intervals on a cross section of a stream.

The Price-type meter is used for higher fluid velocities and deeper stream channels than the Pygmy meter. The Pygmy-meter is limited to velocities below 3 or 4 ft per sec (0.9–1.2 m/s) and is useful for shallow streams. Wading rods or various methods of cable suspension can be used with the Price-type meter. Wading rods must be used with the Pygmy meter. Depth settings for the flow meters, with either suspension method, depend on the water depth. Two fluid velocity measurements are made at each station on a stream cross section if the water depth is 2 ft (0.6 m) or greater. A measurement is made at 0.2 and then at 0.8 of the water depth from the surface. Only one velocity measurement is made if the water depth is less than 2 ft (0.6 m). The measurement then is made at 0.6 of the water depth from the surface (US Bureau of Reclamation, 1975, p. 126).

Stream flow usually is calculated using the mid-section method. This method requires that the fluid velocity be measured at prescribed depths along a number of verticals on the cross section of the stream. The mean of the velocity is multiplied by the total water depth at the vertical and by the sectional width extending halfway to the verticals on each side of the vertical just measured. Flow is calculated as follows:

$$q = \frac{V_1 + V_2}{2}\left[\frac{(L_2 - L_1) + (L_3 - L_2)}{2}\right] d_2 \tag{6}$$

where q is the flow through the section (L^3/T), $V_1 + V_2$ is the fluid velocities at vertical L_2 at prescribed depths of 0.2 and 0.8 or 0.6 of the depth of the water (L/T), L_1, L_2, L_3 are the distances from the initial point for any three consecutive verticals (L), and d_2 is the water depth at vertical L_2 (L) (US Bureau of Reclamation, 1975, p. 127). The total stream flow is the summation of the flow through all the sections.

A third type of meter (propeller) is used for special situations. The propeller meter is less subject to velocity components not parallel to the meter axis and it is also suitable for multiple mounting for simultaneous velocity measurements on a vertical axis.

Stream flow can be measured by tracer velocity or by dilution techniques utilizing a tracer that has been introduced into the stream. Salt can be used for either procedure. The salt-velocity method involves the addition of a slug of salt solution at some point upstream of the section to be monitored. Pairs of electrodes are placed downstream of that point at two locations a known distance apart. Each pair of electrodes is attached to a galvanometer, preferably of the recording type. The time lag between the recorded maximum values measures the time required for the slug of salt solution to pass the electrodes. Stream flow can then can be calculated by:

$$Q = \frac{AL}{T} \tag{7}$$

where Q is the stream flow (L^3/T), A is the cross sectional area of stream (L^2), L is the distance between pairs of electrodes (L), and T is the time measured for the slug of salt solution to travel between the two pairs of electrodes (T) (US Bureau of Reclamation, 1975, pp. 161-162).

The dilution technique for determining stream flow can utilize a number of different solutions or dyes. Salt can be used for this technique also. A known concentration of the salt solution is added to the stream at a constant known rate. The concentration of the solution is determined at some point downstream of the injection point where the solution has become thoroughly mixed with the stream flow. Stream flow can be calculated based on the following relationship:

$$Q = q\frac{C_1 - C_2}{C_2 - C_0} \tag{8}$$

where Q is the stream flow (L^3/T), q is the injection rate for solution (L^3/T), C_0 is the natural or background concentration of solution prior to injection, C_1 is the concentration of injected solution, and C_2 is the concentration of solution, including background concentration, downstream of injection point (US Bureau of Reclamation, 1975, pp. 162-163). Nonhomogeneous stream flow makes this technique less satisfactory than others discussed in this section.

Continuous records of stream flow or spring discharge can be made with stage recorders. The stage recorder must be incorporated into a stilling well associated with a flume, weir, or manually metered section of stream. A semipermanent to permanent installation is required. The stilling well can be installed in the stream channel so that a stage-discharge relationship can be established with any of the periodic measurement techniques described earlier. The stage of the stream is noted each time a flow measurement is made; this relationship can be plotted for interpolating stream flow based on stage measurements only. A simple plot is adequate in most instances (Linsley, et al., 1958, p. 123).

A stilling well and stage recorder can be incorporated directly into some types of flumes so that the stage reading directly indicates stream flow. An intermediate relationship does not have to be established in this case. A stilling well and recorder should normally be installed in the pool formed upstream of a weir installation. The stage record when correlated with the elevation of the weir crest can be used to determine stream flow directly as noted previously.

The maximum stream flow can be recorded with a stage recorder and flume or weir if adequately sized. The stage-discharge relationship method of determining stream flow may not be adequate if a significant extrapolation of the data is needed. Ideally, the maximum flow is measured as it occurs. An indication of the maximum flow can be obtained by measuring the maximum stream stage with a crest stage gage. This gage consists of a length of pipe, the bottom of which has been perforated. The pipe is capped on the bottom and top after emplacing ground up cork and a measuring stick. The rising water level causes the cork in the pipe to rise, whereupon some adheres to the measuring stick at the highest level attained. The maximum stage can be determined by removing the measuring stick and noting the highest level the cork attained (Linsley, et al., 1958, p. 115).

Two water properties are of greatest interest to base flow or spring flow measurements. The temperature of spring water can be used as an indicator of the length of the ground water flow system. A conventional thermometer is an obvious choice but electronic measuring devices also exist. The second water property of interest is the mineralization or total dissolved solids (TDS) content of the water. This also can be an indicator of the length of the flow path. TDS can be determined by having water samples analyzed in a laboratory. Alternatively TDS may be approximated closely by either field or laboratory measurements of the specific electrical conductance (conductivity) of the water. A low conductivity indicates a low total dissolved solids content while a high conductivity indicates a high content of TDS. Natural waters are not simple solutions but the conductivity is easily determined and it provides a practical indication of the total dissolved solids content of the water (Hem, 1970, p. 99). An arithmetic plot of dissolved solids content vs. conductivity appears to be a straight line but there is some curvature to the line. This apparent straight line usually is expressed as:

$$S = kA \qquad (9)$$

where S is the dissolved solids (milligrams per liter), k is the conductivity (micromhos), and A is a conversion factor (Hem, 1970, p. 99). The conversion factor (A) usually is between 0.55 and 0.75, although the full range is approximately 0.54 to 0.96. Higher values generally apply to water high in sulfate concentration (Hem, 1970, p. 99).

Basin Divides

Surface water basin divides usually can be determined from existing topographical maps. Areas that have not been mapped previously or areas requiring greater detail may be mapped using conventional surveying techniques or aerial photographic interpretation.

GROUND WATER SYSTEM

Data collection for the ground water system is the most important aspect of the total data collection effort because these data provide the input from which mine water inflow predictions will be made. Hydrogeologic properties also determine the extent to which surface water systems in the area will be affected by the mine inflow. Conversely, these properties affect the degree to which the surface systems can affect the rate and duration of mine water inflow.

Geology

Conventional Geologic Techniques: Geology can be mapped by conventional field techniques utilizing formation outcrops and other surface features. Particular attention should be placed on determining accurately the location, dip, and strike of such features as strata, faults, fractures, and joints. Geologic structure should be noted or interpreted where possible. Aerial photo interpretation should be used where possible to aid in the delineation of formations and geologic structure; infrared photographs can be useful with the added benefit that ground water discharges, springs, and gaining reaches of streams may be especially visible due to the disparity in temperatures between the ground water and general surface features.

Surface Geophysics: Surface investigative techniques may be extended to include one or all of several basic procedures used in geophysics to delineate subsurface geology. One of the most frequently used methods is based on electrical resistivity. The electrical resistivity of a formation is directly proportional to the length and specific electrical resistance of the material while being inversely proportional to the cross-sectional area of the material. It is this specific resistance governed by the solid matrix and interstitial water chemistry that regulates the gradient in electrical potential that is established by an induced current. Rocks and minerals display a wide range of resistivity unlike any other naturally occurring physical property (Zohdy, et al., 1974, p. 9). Electricity is conducted in most formations by the interstitial fluid. The resistivity is, therefore, controlled more by the porosity, water content, and water quality than by the rock matrix resistivity. Clays, however, can conduct electricity both electronically and electrolytically.

A resistivity survey is conducted by passing a direct or low frequency current into the ground through two current electrodes. The potential drop is measured across two potential electrodes. These four electrodes can be arranged in various configurations and the current and potential measurements can be used to calculate resistivity (Zohdy, et al., 1974, p. 9). Two basic procedures are used in a resistivity survey. These are electrical soundings and horizontal profiling. Electrical soundings are conducted by increasing the spread between the current and potential electrodes. Horizontal profiling is accomplished by moving the entire electrode spread with a fixed spacing along a profile after each measurement. Both procedures sometimes are required so that the influence of horizontal and vertical variations in resistivity can be interpreted for each procedure. Horizontal profiling is used to determine the areal extent of apparent resistivities delineated previously by depth soundings (Zohdy, et al., 1974, p. 13).

Electrical sounding data can be interpreted qualitatively by preparing apparent resistivity maps for given electrode spacings and contouring the results. Alternatively apparent resistivities can be plotted as a section or profile using a given electrode spacing or all spacings. Quantitative evaluations of resistivity data are made by comparing sounding data with published theoretical curves for simple layered geometries (Zohdy, et al., 1974, pp. 33–46).

Very useful information for hydrostratigraphic mapping can be derived from electrical resistivity surveys. This information includes (1) delineation of hydrostratigraphic changes with depth and lateral distance interpreted for the presence of high hydraulic conductivity units important for monitoring, (2) contour maps of marker bed horizons important for structural assessments (such as faults) can be derived from isoresistivity maps, and (3) electrical hydrostratigraphic analysis can point out important target horizons for test well completion. This type of electrical survey may be a cost-effective reconnaissance technique for maximizing the design and location of test holes in the proposed mine area.

Seismic surveys are based on the propagation of artificial pulses of energy through the ground and their detection by electromechanical transducers called geophones. The ground motion induced by the pulses of energy is amplified and recorded with respect to time. Two parameters are of interest in the interpretation of seismic data, the velocity of the pulse and the geometry of the pulse path. A pulse is transmitted by four kinds of elastic waves: only the compressional body wave is of great use for a seismic survey of hydrostratigraphic or formational units in the vicinity of a proposed mine. This type has the highest velocity of the four waves and therefore the shortest travel time for a given path (Zohdy, et al., 1974, pp. 67–68).

The waves from an energy pulse move as a curved front away from the source of energy that is usually created by the detonation of an explosive charge or the impact of a dropped weight. The movement of the pulse as a wave front is best depicted as a ray path that proceeds from the source of the pulse perpendicular or nearly perpendicular to the wave front. One ray path travels horizontally (directly) from the source to a geophone. A second ray path strikes a boundary between formations as would be created by a disparity in their elastic properties. It is actually the discontinuity between formations that is interpreted as to depth and configuration in a seismic survey. Hydrostratigraphic boundaries may be mapped using seismic techniques only if the boundaries are coincident with the changes in elastic properties associated with the formational boundaries. This second ray path strikes the formation boundary at an angle greater than the critical angle and is totally reflected back toward the surface and the geophone. A third ray path strikes the boundary at the critical angle with part of the energy being reflected back to the surface and part refracted, traveling parallel to the boundary. A small pulse is formed in the lowermost part of the unit above the hydrostratigraphic or formational boundary by this part of the ray path traveling parallel to the boundary. It is this small pulse that reaches the surface and is picked up by a geophone. The fourth ray path strikes the boundary at an angle less than the critical angle. Part of the energy is reflected upward and part is refracted down into the lower unit at a new angle that is a function of the pulse velocity in each unit and the angle at which the ray path strikes the hydrostratigraphic or formational boundary (Zohdy, et al., 1974, pp. 68–69).

Two general types of seismic surveys may be useful for hydrostratigraphic mapping. The reflection method employs the second ray path discussed previously. This method yields greater accuracy and resolution than the refraction method; it allows the mapping of a larger number of formations and does not require that each deeper hydrostratigraphic or formational unit have a higher velocity. The reflection method requires that the velocity of the units be known or determined in the field at control points. Continuous profiles between control points are desirable; consequently large crews and sophisticated equipment are necessary. The reflection method is not as well suited for exploration as is the refraction method where steeply dipping boundaries exist.

The second common seismic method is refraction surveying that employs ray paths three and four (above) for detecting a single hydrostratigraphic or formation boundary (discontinuity) and multiple boundaries, respectively. A preknowledge of formation velocities is not required but the velocities must increase between units with depth. Continuous profiles are not necessary to tie in control points for velocity checks; the method is well suited for use in areas with steeply dipping boundaries. The selection of an exploration technique should be based on the expected economic return of the information and the geometry of the hydrostratigraphic or formational units of interest (Zohdy, et al., 1974, pp. 72–73).

General hydrostratigraphic trends and features may be detected by a gravity survey (gravimetry). This technique requires accurate measurement of the force of gravity at locations within and adjacent to the area of interest. Variations in the measured force of gravity, corrected for factors such as elevation and latitude, and compared with the gravity field of a reference spheroid, reflect variations in the density of rock in the area. The force of gravity is measured with a gravity pendulum or more commonly a gravimeter. The gravimeter is a highly sensitive spring balance that allows the measurement of the weight of an internal mass. Differences in the weight of the internal mass reflect variations in the acceleration due to gravity since the mass remains constant (Zohdy, et al., 1974, pp. 85–86). Interpretation of the gravity data requires several corrections as noted plus a knowledge of relative rock densities in the area surveyed. A solution for a given gravity anomaly is never unique and the quality of the interpretation is also a function of the kind and amount of geologic information available (Zohdy, et al., 1974, p. 97).

Additional geophysical investigations for hydrogeological reconnaissance can be conducted by measuring the direction, gradient, or intensity of the earth's magnetic field. Variations in these quantities are interpreted for their geologic significance. The magnetic intensity can be measured with simple or elaborate instruments from the land surface or aircraft. Variations in the earth's magnetic field or anomalies are caused by the presence of magnetic material near the surface. The character of an anomaly may be interpreted as to the general rock type while a quantitative evaluation can yield information on the structure, extent, and depth of burial of formational units and their magnetic properties. The magnetic anomaly maps can also illustrate the depth and extent of hydrostratigraphic units if they have mappable magnetic properties. One example of this hydrostratigraphic unit mapping is in aquifer delineation within basalts (Lankston and Lankston, 1977). Most sedimentary rocks are essentially nonmagnetic and not well suited to magnetic surveys. Igneous and metamorphic rocks are better suited for such a survey because of their greater content of magnetic minerals (Zohdy, et al., 1974, p. 107). Magnetic data must be compensated for diurnal fluctuations in the magnetic field before presenting the data as profiles or contour maps.

Test Drilling: Subsurface hydrogeology can be investigated by controlled test drilling operations. Numerous drilling methods exist and the selection of a particular method is dependent upon several factors including site access, type of hydrostratigraphic units anticipated, projected depth, type of samples required, and the desired angle of penetration. Formations can be identified and located by evaluating drill cuttings or samples vs. depth of penetration. The drilling rate indicates the relative hardness of the formations encountered based on an evaluation of the cutting capability of the bit, drill stem weight, and rotational velocity for rotary drilling methods. Excellent formation samples can be obtained via coring, a standard exploration technique. Core samples can be utilized for formation identification and also for the determination of various hydrogeologic properties that will be discussed in the following section.

Some drilling methods utilize a drilling fluid to lubricate the bit and provide a medium for the removal of drill cuttings from the bit face. These fluids operate on a continuous closed circuit so that fluid must be added to the drill hole as the depth of the drill hole increases. Fluid losses occur when high hydraulic conductivity units are encountered with a low potential energy level. Conversely, excess fluids will occur if a high hydraulic conductivity unit is encountered with a potential energy level above land surface. Both situations yield a relative value on the distribution of water levels vertically and on the relative hydraulic conductivity of the units penetrated.

Borehole Geophysics: Borehole geophysics can be used to aid in the delineation of formations and as an evaluative tool with regard to hydrogeologic properties. Williams (1982) discusses the use of borehole geophysical logs for purposes of delineating hydrostratigraphic units as opposed to geologic units. Borehole geophysics consists of lowering various sensing devices into a borehole so that physical parameters can be measured and recorded. The recorded data are referred to as a geophysical well log. These logs can be evaluated and interpreted to determine the lithology, geometry, resistivity, formation factor, bulk density, porosity, estimated hydraulic conductivity, moisture content, and estimated specific yield of hydrostratigraphic units, and to estimate the source, movement, and chemical and physical characteristics of ground water (Keys and McCary, 1976, p. 2). One of the most important aspects of geophysical logs is the fact that one is able to assess continuous in-situ variations in subsurface borehole properties due to the continuous nature of the log. This is not possible with chip samples that usually are mixed and taken only at selected intervals and cores that generally do not have good recovery in porous incompetent materials.

Direct access to the subsurface can be achieved directly by mining or tunneling or indirectly by drill holes. Drilling is expensive and it is important that the maximum amount of information be derived from a drill hole. Hydrostratigraphic units can be defined from interpreted geophysical well logs. These logs can reduce future drilling costs by guiding the construction of future test holes. The vertical and horizontal extrapolation of borehole data is very important when assessing the degree of continuity of high hydraulic conductivity hydrostratigraphic units that constitute preferential flow paths for ground water.

The hardware for borehole logging consists of similar basic components for all the different tools consisting of a sensor, signal conditioners, and a recorder. The sensor or sonde receives power and transmits the signal to the surface through a conducting cable that positions the tool in the hole by means of a winch. Logging speed and the direction of the sonde movement are controlled at the surface. Power to the sonde, signal conditioning, and recorder responses are also handled at the surface. The return signal from the sonde is a function of lithologic fluid and borehole parameters and usually is displayed on chart paper and stored on magnetic tape (Keys and McCary, 1976, p. 21).

The electronics for borehole sondes has been

miniaturized so that the probes are as small as 1 in. (25.4 mm) in diameter; most sondes are not greater than 4 ft (1.2 m) in length. This increases their versatility for use in small diameter exploration holes. Tools that have the capability of recording more than one parameter, sometimes called combination logs, may be of larger diameter and length. It commonly is more expedient to run single sondes within the borehole at slow logging speeds for small-scale engineering parameters. The maximum amount of information can be obtained in this way. High logging speeds decrease resolution of bed boundaries that limit the effectiveness of the logs.

A composite analysis of many different geophysical logs will yield more reliable information than interpretations based on only one log response since each log measures a different physical parameter. A composite analysis will tend to support or contradict conclusions drawn from other logs. A minimum of three logs provides greatest assurance for correct interpretations. Similarly, it is statistically more meaningful to interpret geophysical logs from a number of wells in one geologic environment (Keys and McCary, 1976, p. 15).

Nuclear Logs—Nuclear or radiation logs measure radiation emitted from the nucleus of an atom. They have a distinct advantage over most other geophysical logs because they can derive information from cased or open holes that are filled with any type of fluid (Keys and McCary, 1976, p. 58). The most common nuclear logs are natural gamma, gamma gamma, and neutron epithermal neutron. The primary functions of these logs are to identify variations in lithology, density, and moisture content or porosity within the borehole materials.

The record of the natural gamma radiation emitted by borehole materials is called a natural gamma log. This log is used primarily for the identification of lithology, especially clay units. These logs are useful for hydrostratigraphic correlation in open or cased drill holes that can be liquid or air-filled (Keys and McCary, 1976, p. 64). Heavy radioactive elements generally are concentrated in clays through processes of ion exchange and adsorption. Consequently these units generally show high gamma intensities and often make good stratigraphic markers for regional correlation. The natural gamma log is used in conjunction with other radiation logs to delineate the hydrostratigraphy of

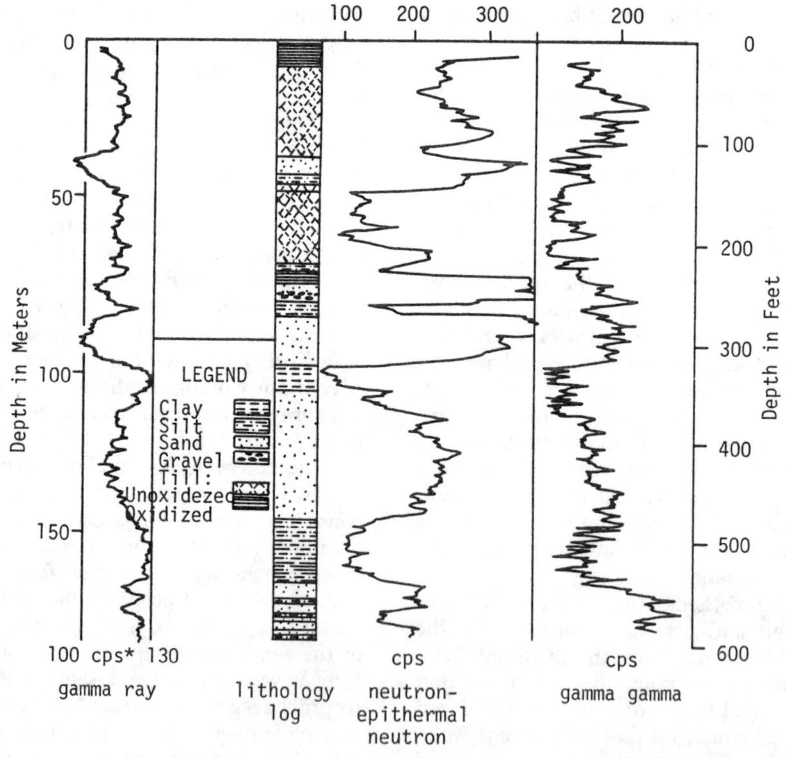

* counts per second

Fig. 18. Delineation of lithology with geophysical logs in an uncased hole (modified after Dyck, et al., 1972).

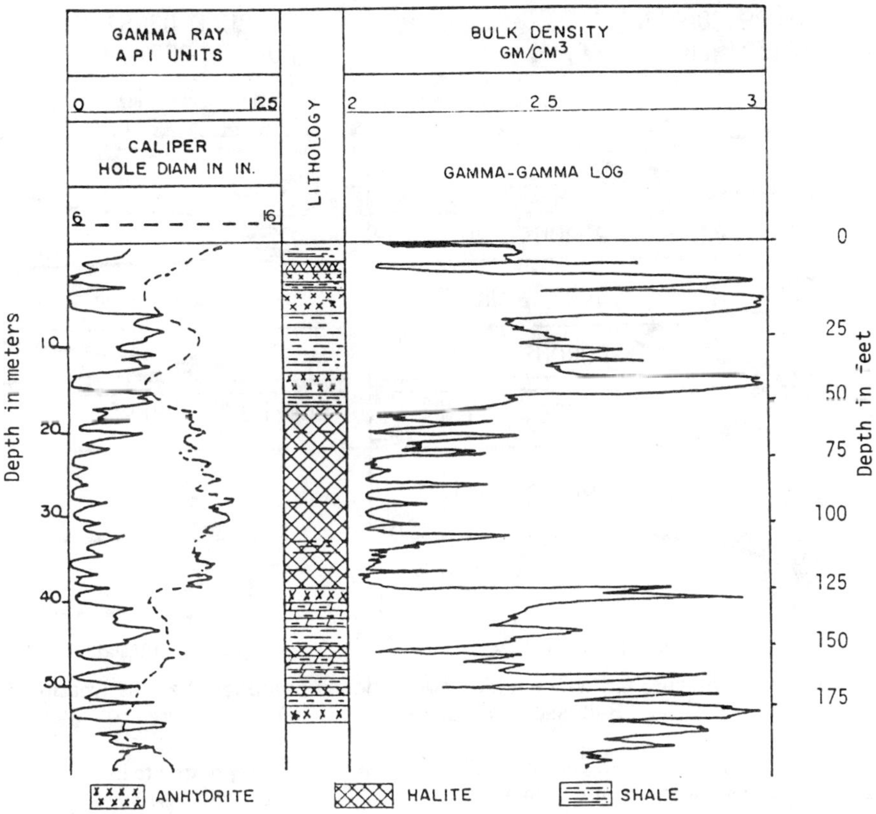

Fig. 19. Varying densities delineated by the gamma gamma log (modified after Tixier and Alger, 1970).

a section (Fig. 18). The presence or absence of clay units of homogeneous nature is important in assessing a site for the potential vertical movement of ground water due to the clay's lower hydraulic conductivity. Consequently, the natural gamma log is important in measuring clay content, homogeneity of clay layers, and lateral continuity of clay zones.

The gamma gamma log is an induced radiation tool that records gamma photons that have been back-scattered and attenuated within the borehole and surrounding rocks. These logs aid in the identification of lithology and provide a means of measuring the bulk density and porosity of the formation (Keys and McCary, 1976, p. 70). Gamma photons undergo Compton scattering with orbital electrons from the borehole environment. Due to this photon scattering and subsequent lowering of energy levels, less radiation is detected by the sensor in media of high electron density. Conversely, higher back-radiation is detected in high porosity zones. This relationship is used to delineate lithologies of different densities which affect subsurface permeabilities. Fig. 19 illustrates the use of the gamma gamma log to delineate hydrostratigraphic units of varying density.

This tool that measures continuous variations in bulk density is valuable in assessing the locations of permeable hydrostratigraphic units. Unfortunately, the gamma gamma log has a small radius of investigation (zone from which information is derived due to photon return) so the tool is best suited for use in small diameter boreholes where caving has not been a problem. Borehole rugosity affects the responses of all radiation tools due to the irregularity of material in the zone of investigation—borehole fluid vs. surrounding wall rock. The gamma gamma log is most susceptible to this rugosity factor and should be run decentralized within the borehole so that the source and detector are deriving information from the wall rock and not just the borehole fluid medium.

The configuration of the probe in neutron logging is designed so that the tool responds primarily to a function of hydrogen content in the borehole and surrounding lithology. Neutron energy is modified by elastic collisions with borehole elements and the effectiveness of these energy moderators is

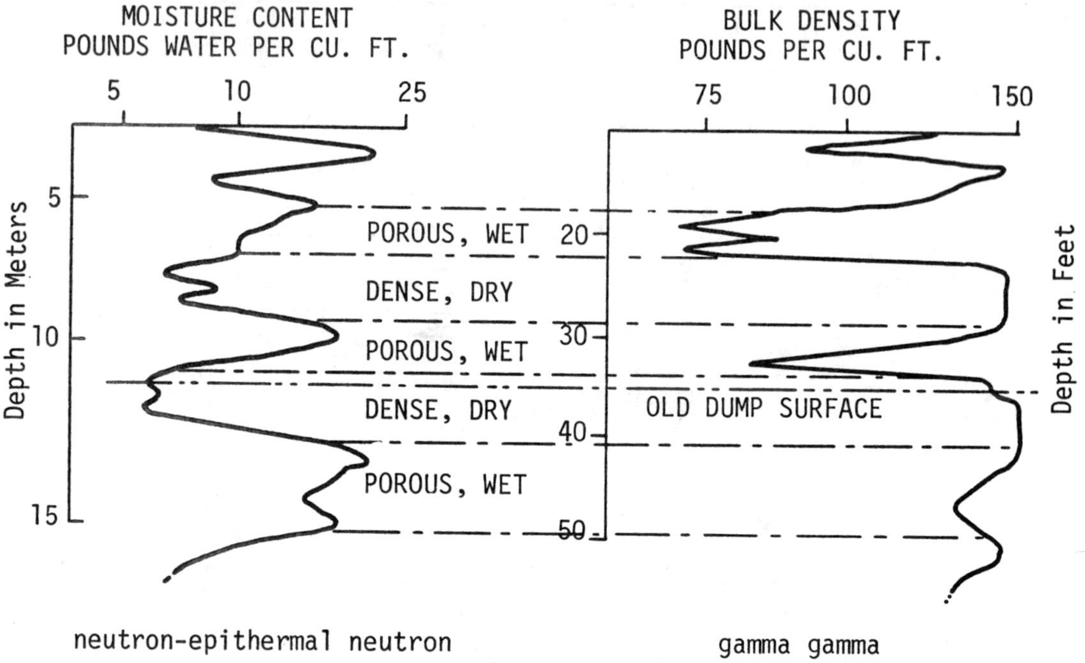

Fig. 20. Delineation of perched water tables above dense compacted zones (modified after Baltosser and Lawrence, 1970).

a primary factor in neutron logging. The mass of the nucleus of a hydrogen atom has approximately the same mass as the neutron, so a collision of a neutron with a hydrogen atom causes the neutron maximum energy loss. Consequently, the energy of the neutron response is interpreted for moisture content in the unsaturated zone of the borehole (above the water table) and total porosity below the water table (Keys and McCary, 1976, p. 74). Neutron logs can show the location of porous hydrostratigraphic units, perched water tables, and dense ground water confining zones (Fig. 20).

The neutron epithermal neutron log is commonly a key indicator for the presence of relatively high hydraulic conductivity hydrostratigraphic units within the borehole. This log has a relatively large radius of investigation and is very sensitive to variations in moisture and porosity. When interpreted with respect to other radiation logs and a caliper log, the neutron log can indicate variations in grain size distribution and the presence of fractures that may have otherwise gone undetected (Fig. 21).

Fig. 22 depicts the degree of detailed information available from radiation logs in a borehole exhibiting subtle but important changes in grain size distribution. The neutron-neutron log is particularly useful in correlating samples with log response variations. This correlation, supported by the other radiation logs, makes initial assessments of porosity and approximate hydraulic conductivity possible. An overall assessment of the section's characteristics has been accomplished by determining the percentage by type of various hydrostratigraphic units in the borehole. Tables 2A and 2B are a partial display of this assessment.

The results of all of the radiation logs are based upon a statistical return response of the photons and neutrons that is a function of logging speed. When the objective of the logging program is to delineate potentially small-scale hydrostratigraphic units that may dominate ground water flow systems, it is important that the logging speed be slowed to allow for a statistically significant return of radiation to the detector. A maximum logging speed of 15 fpm (0.8 m/s) is desirable to yield the degree of detail necessary for this type of assessment. Faster logging speeds tend to give an average response over a larger zone of the borehole that masks individual hydrostratigraphic units and consequently decreases the effectiveness of the logging technique.

Fluid and Mechanical Logs—In addition to the lithologic parameters, a second important category of information that can be derived from well logs includes data on the character and movement of water in the borehole and the formation. Characteristics of well bore fluids can be measured relatively easily via direct sampling, but data on formation fluids are more difficult to obtain; some

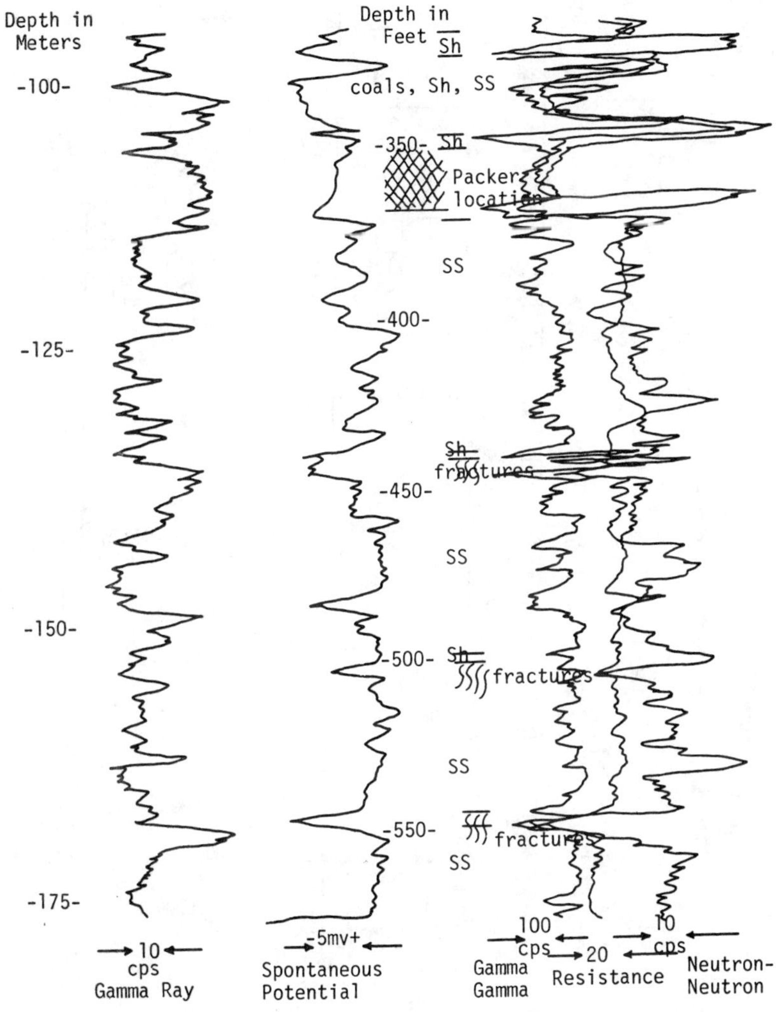

Fig. 21. Radiation logs indicating the presence of fractures and lithology interfaces. Note the variations in sandstone between 134 and 164 m (440 and 539 ft) (after Williams, 1982).

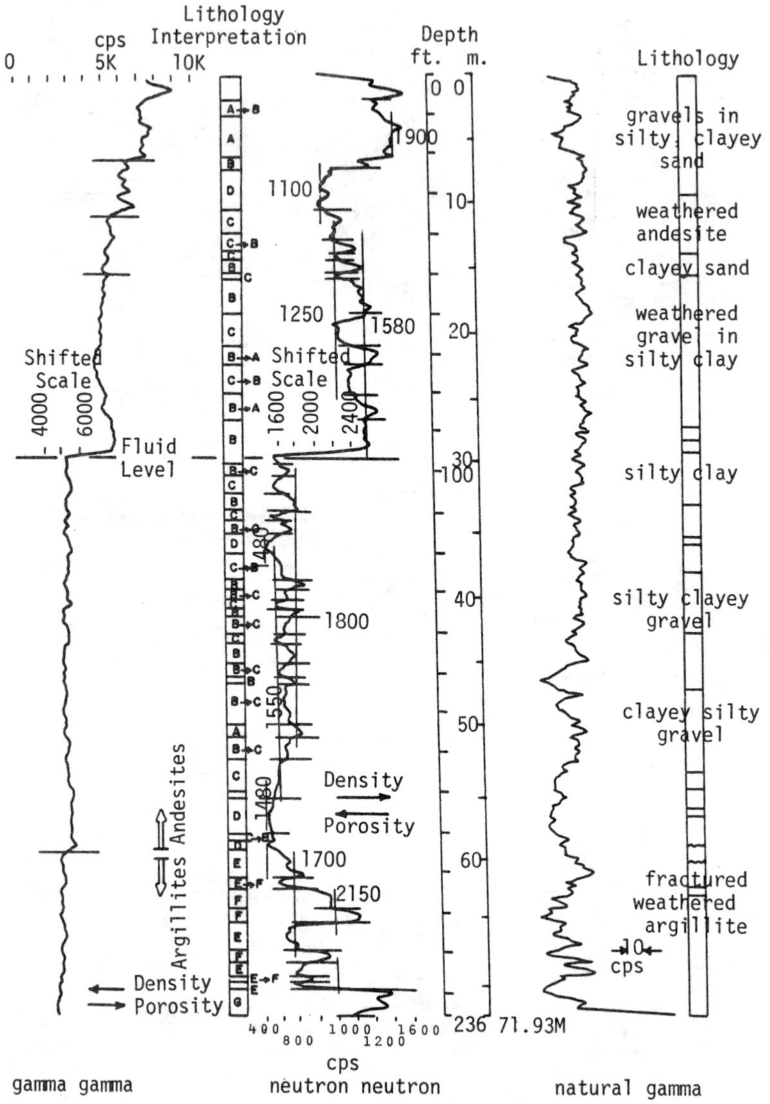

Fig. 22. Partial display of radiation logs delineating hydrostratigraphy based on grain size and weathering variations (after Williams, 1982).

Table 2A. Correlation of Geophysical Log Response with Borehole Characteristics (Williams, 1982)

Geophysical response type (andesites)	Description	NN count range (*above water table)	Estimated porosity range	Estimated† relative permeabilities
A	Predominately gravels and sand in silt matrix, weathering, more sandy than silt, small percent clay.	1900*	10–15%	M, H-M
B	Gravelly, weathered gravels in silty/clay matrix, weathering, friable	1580* 1800	15–20%	M-L, M
C	Sandy with few gravels in predominately clayey matrix, some silts	1250* 1550	20–30%	L, M-L
D	Stiff, sticky clay, small percent silt, possibly very few gravels included	1100* 1480	30–35%	S

† Estimated relative hydraulic conductivities: H = high, H-M = high to medium, M = medium, M-L = medium to low, L = low, L–S = low to slight, S = slight.

characteristics may be interpreted from geophysical logs. Logs that measure fluid parameters are important in the assessment of subsurface hydrostratigraphy because the logs can indicate the locations of zones with varying water chemistries. Mechanical logs can indicate the relative direction of movement of water within the borehole, thereby providing an indication of hydraulic potentials within or between hydrostratigraphic units.

The caliper log is a record of the irregularity and the diameter of the borehole. This log permits the evaluation of the rugosity of the borehole in which other logs are made in order to correct them for hole diameter variations. The caliper log can also be used to provide information on lithology by identifying zones of incompetence where breakouts occur.

Temperature logs record the thermal gradient of the borehole fluid by use of a single sensor. The fluid temperature may be representative of the thermal conductivity of the rock if the fluid in the well is in thermal equilibrium with the adjacent rocks and if there is no vertical circulation of fluid in or near the borehole (Keys and McCary, 1976, p. 99). Vertical movement of the borehole fluid causes mixing of water at varying temperatures that appears as a homogeneous water temperature on the log. Conversely, influxes or effluxes of water can cause rapid temperature changes at specific zones within the borehole. These zones can be interpreted with respect to the location of hydrostratigraphic units of varying hydraulic heads (Fig. 23).

Geophysical logs that indicate fluid flow are useful for the determination of relative hydraulic potential distribution in a well, much like the temperature log. Identification of recharge and discharge areas plus hydraulic head changes among hydrostratigraphic units are all important parameters. In-hole flow commonly is measured with an impeller flowmeter designed to measure vertical movement. Vertical mixing of waters from different hydrostratigraphic units in a test hole is important in determining whether water samples are to be taken for water quality analysis. The direction of fluid movement indicates varying hydraulic heads within or between hydrostratigraphic units and gives an indication of approximate, relative hydraulic conductivities (Fig. 23).

In very shallow holes [less than 100 ft deep (30.5 m)] or in holes that are completely cased with no openings, the fluid and mechanical logs may not

Table 2B. Percent Type Within Borehole (Williams, 1982)

Type	Total footage	% by type*	Estimated porosity	Estimated relative hydraulic conductivity
A	12	5	10–15	M, H-M
B	52	23	15–20	M-L, M
C	40	17	20–30	L, M-L
D	25	11	30–40	S

* (Total footage by type ÷ total section) 100.

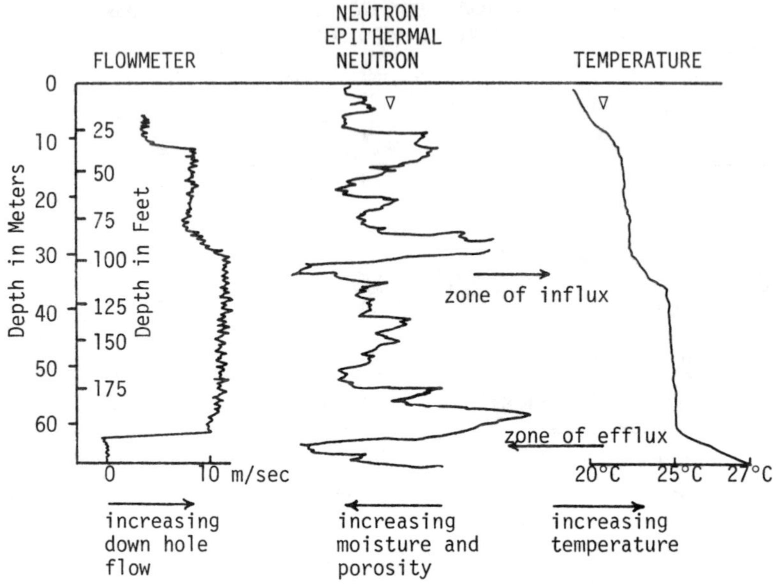

Fig. 23. Borehole logs indicating aquifer potentials (Williams, 1982).

be useful. This statement is based on the fact that a significant geothermal gradient cannot be established within the top 100 ft (30.5 m) of the subsurface and temperature anomalies usually are not significant enough to allow for flow system identification with the temperature log. A completely cased borehole does not allow interaction between hydrostratigraphic units, so no vertical circulation that would be indicative of differing hydraulic heads would exist within the hole. In general, when a decision is necessary concerning which fluid or mechanical logs should be run, the caliper log has the highest priority because of its importance in the interpretation of other logs, with the temperature log a second priority. Fluid flow logs are run only in open boreholes or boreholes with casing perforated in more than one unit. It should be noted also that the caliper log is not useful in delineating borehole rugosity if the borehole is completely cased. It is best to conduct all geophysical logging in an open borehole if possible; this practice maximizes the sonde responses and allows for the greatest freedom in selecting tools to be run. A caliper log is a must in all open holes when radiation logs are to be interpreted.

Electric Logs—Electric logs provide information about lithology, hydrostratigraphic unit thickness, fracture location, and fluid movement in the borehole. Many types of resistivity and resistance tools are in use; only the single point resistance log is discussed here because its design lends itself to engineering test holes for lithologic investigations.

Electric logs are a measure of natural electrical potentials and resistivities. These logs can be run only in uncased holes that are filled with a fluid such as drilling mud or water. There are many types of electric logs that can be run in a borehole, the choice of which is dependent upon the purpose of the survey. The spontaneous potential log is a record of the small differences in voltage that develop at the contacts between the borehole fluid, borehole lithology, and formation fluid (Keys and McCary, 1976, p. 24). Other electric logs include the resistance and resistivity logs of which there are many types; varieties include single point and differential resistance, short normal, long normal, lateral, microlog, microfocus log, and the guard or lateral resistivity log. Each of these devices has a specific application, depending upon the lithology, depth of drilling mud invasion, and area of interest within the borehole. It is not necessary to run all the electric logs to determine variations in lithology, aquifer locations, and relative conductivities of porous media fluids. The petroleum industry has developed electric logging techniques for use in evaluating the invaded zone since such holes are drilled with muds under high pressure. Since this is commonly not the case in a hydrogeologic test hole, logs such as the log normal and lateral devices usually are not needed. For more information on the various types of resistivity logs, it is suggested that the reader refer to publications by Guyod (1966) and Pirson (1963).

Spontaneous potential (SP) logs measure the small differences in voltage that develop in the borehole as a result of fluid movement and electrical potentials due to borehole fluid migration at lithologic contacts. The tool is similar to a voltmeter that measures changes in potential as it moves within the borehole. Spontaneous potential logs can be very useful in showing fault zones and water-producing units in the borehole (Fig. 24). Streaming potentials are caused by electrokinetic phenomena associated with the movement of an electrolyte through a permeable hydrostratigraphic unit. This electromotive force appears in the borehole at places where mud is being forced into the permeable hydrostratigraphic unit although streaming potentials may be generated in zones of gaining or losing water (Keys and McCary, 1976, p. 24). This SP log is also useful in delineating relative hydraulic conductivities based on the knowledge of borehole fluid potentials. When the water within a hydrostratigraphic unit is much more conductive than the borehole water, the current follows a path from the borehole into that unit. An upward moving electrode senses a decreasing potential because the current flows parallel to the well bore. This relationship would be reversed in the case of good quality water within a hydrostratigraphic unit and a positive SP response would appear opposite the hydrostratigraphic unit (Keys and McCary, 1976, p. 25). Consequently, the negative or positive response on the SP log can be used to infer relative formation water qualities if the fluid conductivities are known.

Single point resistance logs measure the resistance to current flow of the borehole fluid and material. The logs are useful for geologic correlation because of their unique response to changes of lithology. Fractures or zones of weakness along with changes in borehole fluid conductivities can usually

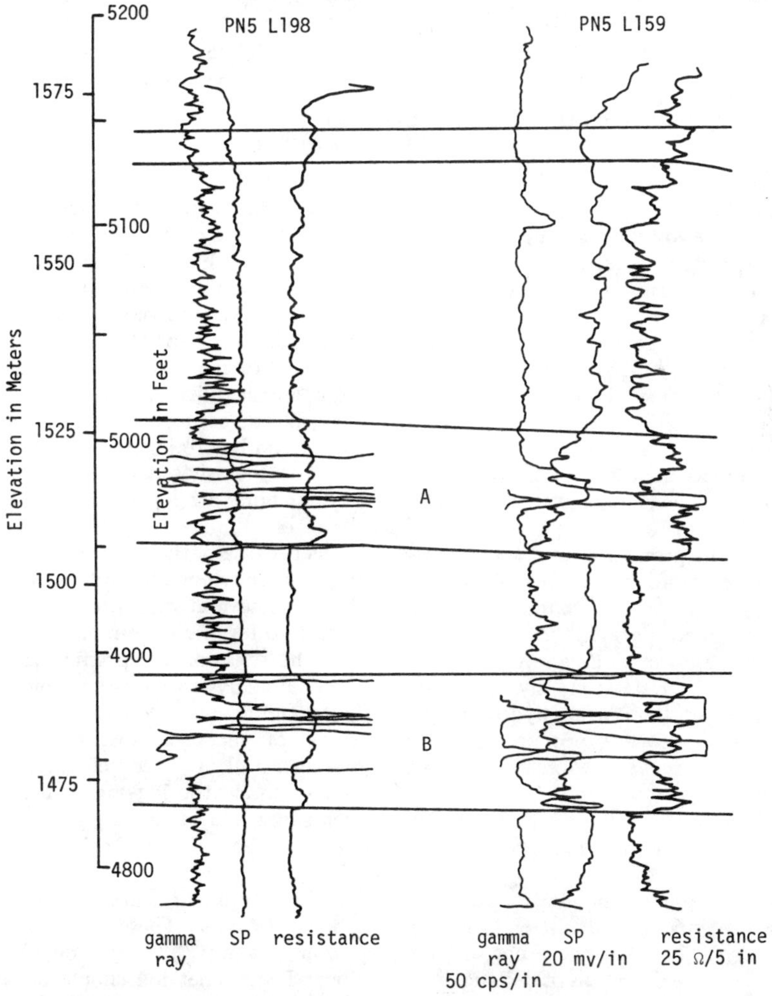

Fig. 24. Correlation of aquifer zones A and B using electric gamma ray logs for uranium in situ leach assessments (after Williams, 1982).

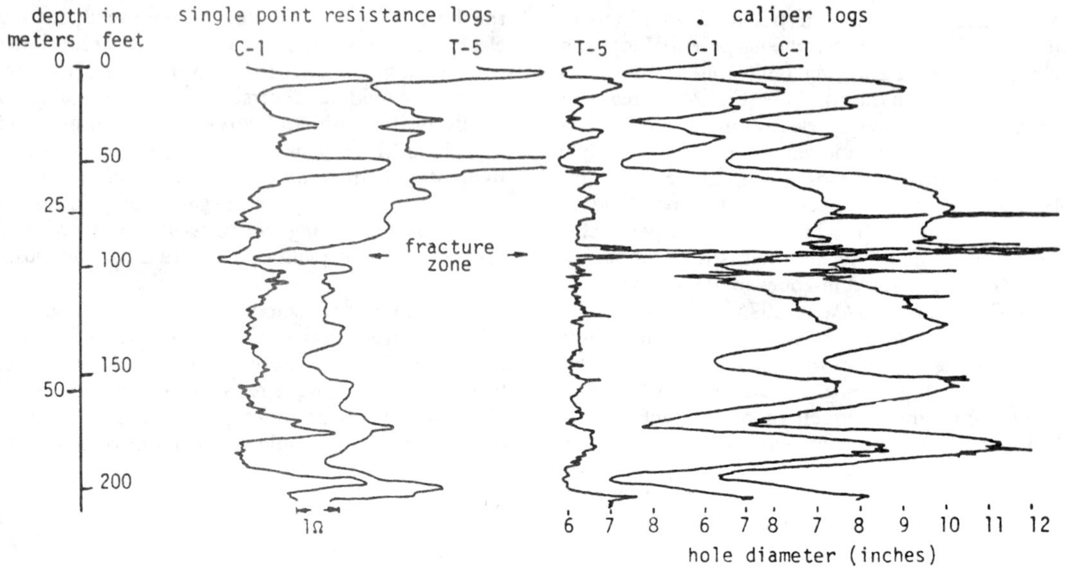

Fig. 25. Comparison of fracture mapping with caliper and single point resistance logs (modified after Keys and McCary, 1976).

be mapped with resistance logs (Fig. 25). For these reasons, the resistance log is sometimes used as a sensitive caliper log and as a relative water quality indicator. The SP and resistance logs are usually interpreted as a unit.

Several other types of geophysical logs that are not discussed herein are available. They are omitted because they are not as useful for the delineation of hydrostratigraphic units. The reader is referred to general texts or journals such as *Log Analyst.*

The number of boreholes required at a mine site is dependent upon the complexity of the hydrostratigraphy of the site. The overall objective of geophysical logs is to provide a continuous sampling of subsurface variations in density and porosity that may be indicative of ground water flow units. Once the hydrostratigraphy has been delineated, further logging would determine whether structure offsets the hydrostratigraphy or in any way influences the overall subsurface hydrogeologic characteristics. The optimal procedure for drill holes whose primary purpose is geophysical logging includes the following:

1) Collect a continuous core or extract samples systematically throughout the length of the borehole.

2) Maintain a maximum and constant hole diameter of 4 in. (10 cm). This diameter facilitates entry of all geophysical probes but is not so large a diameter as to decrease the radius of investigation.

3) Obtain logs in an open hole without casing.

4) Whenever possible, remove the drilling fluid from within the borehole before logging commences. Frequently, however, this will not be possible. In these cases refer to procedure 5.

5) If the hole will not stand open with fresh water within the borehole, it is better to log through drilling mud or a biodegradable type of drilling fluid additive than to log through casing.

6) If casing is required during drilling because of hole instability, then the hole preferably should have plastic casing installed that is cemented to the formation before logging.

The tool should be decentralized if large diameter holes are logged. It is best for the holes to be of one diameter rather than containing reductions because of the difficulties in the mechanics of logging and log interpretations in multidiameter holes. If several reductions occur it may be necessary to log the hole in stages.

The selection of specific logs for hydrostratigraphic analysis at a given mine site is dependent upon the general geology at the site. A site composed of unconsolidated sediments that requires casing in all of the boreholes would not be conducive to electric logging or fluid-flow logging as compared to a site in volcanics that would hold an open borehole and yield substantial information via electric logging.

In general, the suite of radiation logs always should be run. These are the natural gamma, gamma gamma, and neutron epithermal neutron logs. The caliper log should always be run with radiation tools unless the logs are taken entirely within casing, at which point the caliper log is not able to give borehole rugosity information. However the caliper is useful in examining the casing for pinching, swelling, or broken areas.

Hydrogeology

Potentiometric Distribution: The potentiometric distribution (hydraulic head distribution) can be determined with a variety of different types of observation wells. The simplest installation consists of a single-casing well open at a specific position in a hydrostratigraphic unit as illustrated in Fig. 26. This installation allows the measurement of the static water level in only one hydrostratigraphic unit that is unit 3 in Fig. 26. The vertical, potentiometric distribution between or among units can be determined by similar installations in each unit with this type of observation well (piezometer). Additional observation wells at other locations would be required to determine the potentiometric distribution laterally for each hydrostratigraphic unit.

Multiple depth observation wells utilize a single borehole for determining the vertical potentiometric distribution at a specific site. Three basic types of multiple depth observation wells are utilized for this purpose. One type consists of multiple, small diameter, access tubes, each of which has been positioned and isolated so as to monitor only a prescribed position in each hydrostratigraphic unit (Fig. 27). Obvious difficulties can arise in identifying the proper measuring tube with its associated hydrostratigraphic unit after the installation is completed. Sealing the borehole between units is difficult, especially as the number of tubes increases. The material used to backfill the borehole around the tubes tends to bridge, preventing the proper sealing of lower units. Artesian conditions can also complicate the sealing and backfilling operations. In order to eliminate this problem it is necessary to seal from the bottom up with a tremie line.

The potentiometric distribution can also be determined vertically at a site by utilizing a staged casing installation in a single borehole as shown in Fig. 28. A large diameter hole is drilled to or near the contact between units 1 and 2. The large diameter casing may be perforated prior to installation in the borehole. The bottom of the casing is grouted in place with care being taken to prevent sealing the perforations. The grout in the bottom of the casing is then reamed out and the borehole extended to near the contact between units 2 and 3. Grout is again placed in the bottom of the borehole to seal the second string of casing in the borehole. The perforations can be manufactured prior to installation or developed in place depending upon the size of the casing and conditions in the borehole. A packer (grout basket, etc.) is installed on the second string of casing above its perforation zone so that the annular space between the strings of casing is sealed. This procedure makes it possible to use the annular space between the first and second strings of casing as an observation well for unit 1. The grout in the bottom of the second string of casing can be reamed out and the drill hole extended to the contact between units 3 and 4. A third string of casing can be installed. This string can be grouted in place and perforated as was the second string of casing. Successive strings of casing can be installed in this manner until the final inner casing is installed. This string serves as an observation well for the deepest unit (unit 4 in Fig. 28), once this string has been grouted in place.

The installation of a staged casing observation well can be expensive and time consuming due to the number of large diameter casings involved and the nature of the multiple drilling and grouting operations required. Care must be taken to insure proper sealing of the annular space between casings without sealing the perforations or the zone that should be monitored. The inadvertent placement of grout in the perforations and the annular space between the casing and the borehole can sometimes be remedied by in-place shot perforation techniques. This method of installation may be required where squeezing formations exist. Drilling may not be able to proceed beyond a squeezing formation without casing off the formation.

A recent development in this field utilizes a single borehole and casing wherein numerous hydrostratigraphic units can be monitored individually. A continuous borehole is drilled to the base of the lowest unit or position to be monitored. The positions of the monitoring ports and packers, as shown in Fig. 29a, are determined from a careful evaluation of the drilling log and geophysical logs for this borehole. Inflatable packers are generally used for isolating units or positions within a given unit, although bentonite or grout seals can be used also. The monitoring ports are positioned so that specific hydrostratigraphic units or positions within a unit can be monitored. The monitoring ports consist of either a ball valve that can be opened with a pressure sensing tool that forms a seal around the port or a sliding valve (Figs. 29b and 29c) as developed by Patton (1979). A separate tool is required to position the sliding valve so that the ports are open. In both instances, only one port is opened at a time so that the measured water level or equivalent pressure is representative of a specific position in a hydrostratigraphic unit. Multidepth observation wells can be located at various sites around the mine area so that both the vertical and lateral potentiometric distribution can be determined and monitored during the life of the mine.

Water levels in these observation wells can be measured and monitored by several techniques. The customary procedure is to measure the depth to

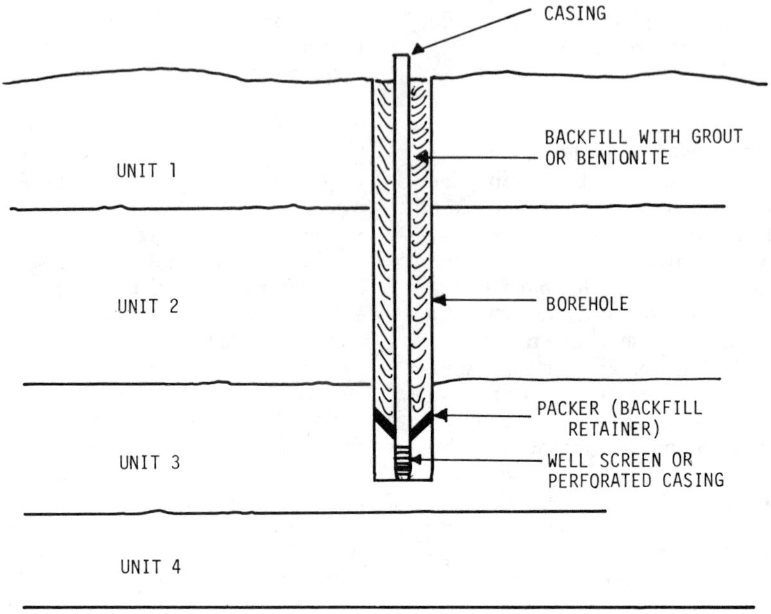

Fig. 26. Single casing, one position monitoring well.

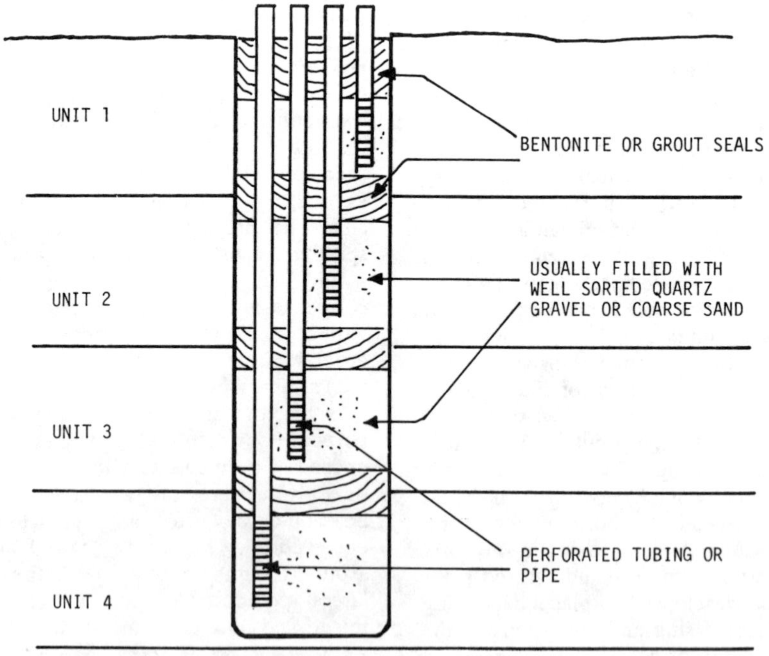

Fig. 27. Multiple depth observation well utilizing several small access tubes.

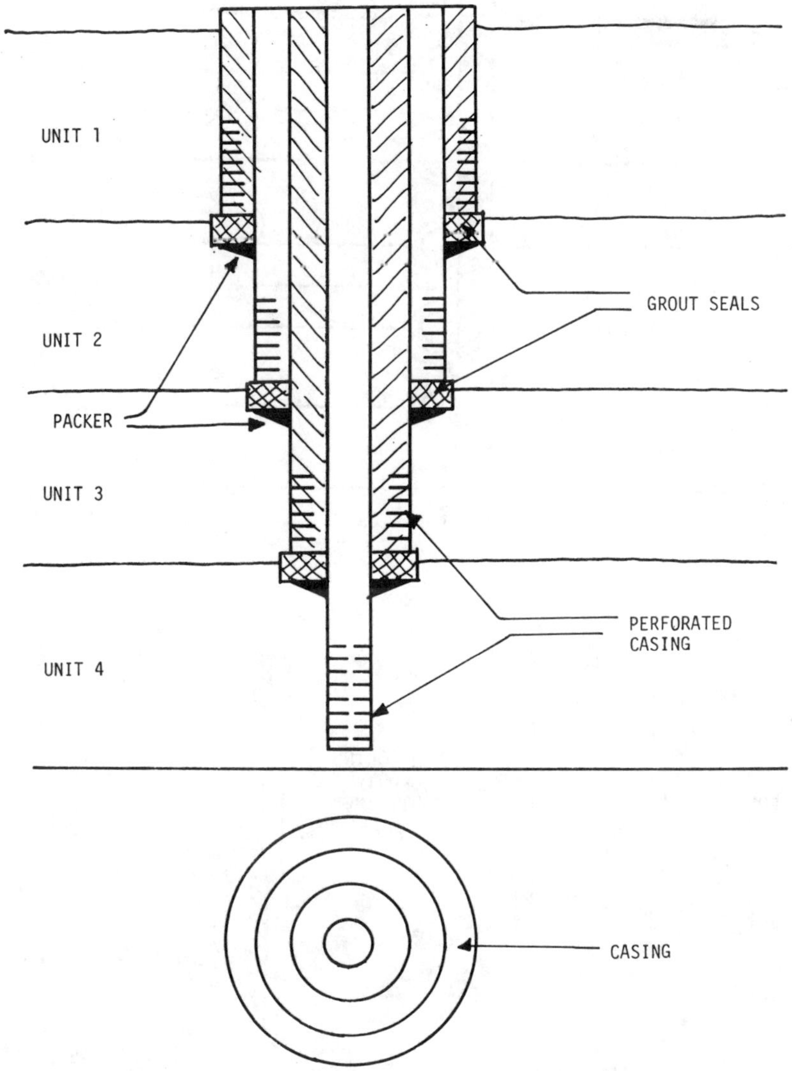

Fig. 28. Multiple depth observation well utilizing a staged casing.

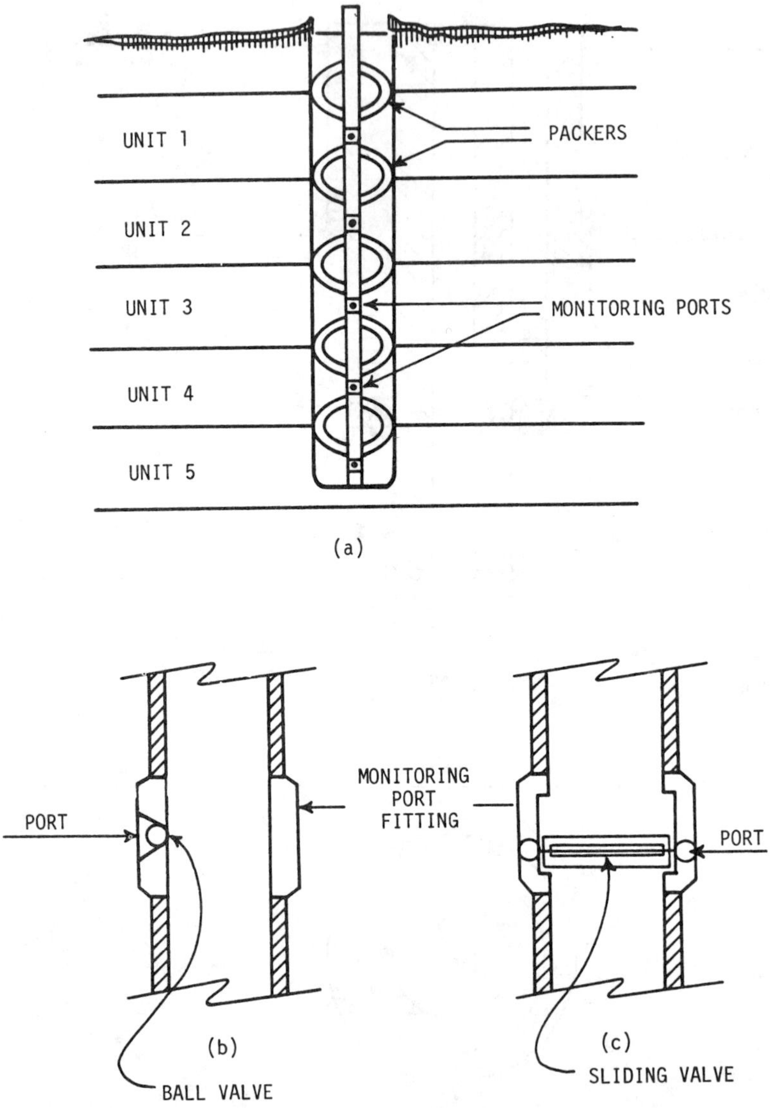

Fig. 29. Multi-depth observation well utilizing single casing.

water in a well from some reference point of known elevation (usually a point on the casing). The elevation of the static water level is calculated easily. An accurate procedure for measuring the depth to water entails the use of a piece of carpenter's chalk, a surveyor's steel measuring tape, and a modest weight that can be attached to the end of the tape. The lower five or ten feet of the tape are dried and chalked prior to lowering the tape in the well. Enough sound is usually produced by the weight striking the surface of the water that the lowering of the tape can be stopped and a distance marker on the tape selected. This marker is lowered carefully to the measuring point and then raised. Subtracting the wetted portion of the tape from the distance held at the measuring point yields the depth to water. Measurements made in this manner should be accurate to about 0.005 foot (Davis, et al., 1975, p. 55). Accuracy of the depth to water measurement is dependent upon the operator's technique and the vertical alignment of the well. Other techniques or equipment should be utilized where cascading water occurs in the well or where rapid measurements must be made.

Another common technique used to measure water levels involves the use of an electric measuring tape referred to as an M-scope, electric tape, or electric sounder. This device is usually rigged as a two-wire cable on a spool, marked in 5-ft, 10-ft, or 1-m intervals measured from a probe. A complete electrical circuit is formed when the probe is sufficiently immersed in the water. Either a light, buzzer, or small ammeter is observed to detect when the probe is immersed since this completes the electrical circuit; the cable is held at the measuring point when this occurs. The distance from this point to one of the marked intervals is noted and the depth to water calculated. Accuracy is generally to within 0.1 ft but readings can be nearly as accurate as the surveyors steel measuring tape if the marked intervals on the cable are calibrated carefully. This technique frequently can be used where cascading water exists in the well if the probe is shielded adequately.

The air line method of determining depths to water in wells requires a pressure gage, pressurized gas source (such as a tire pump or air tank), tubing, and the appropriate fittings to allow an airtight

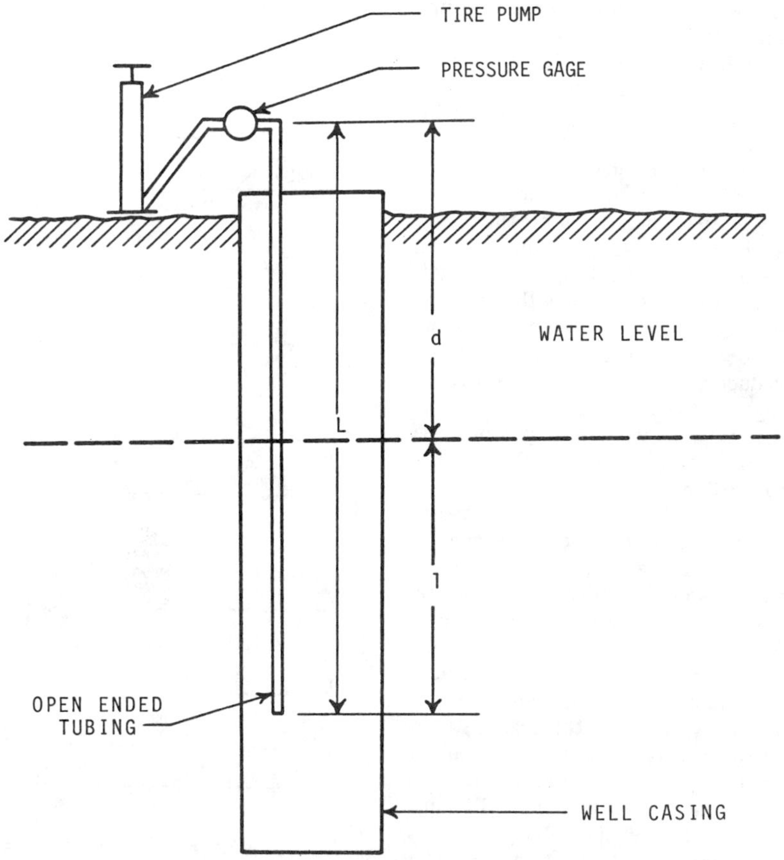

Fig. 30. Air line method of determining depths to water.

connection between the parts. A measured length of tubing is hung vertically in the well; the upper end is attached to the pressure gage and the pressurized gas source as illustrated in Fig. 30. Air or some other pressurized gas is applied to the gage and tubing until a maximum pressure is reached. Gas is bubbling out the end of the tubing at this point. Depth to water is calculated as follows:

$$d = L - l \tag{10}$$

where d is the depth to water (L), L is the depth to bottom of airline (L), and l is the pressure head on gage equivalent to the submerged length of airline (L) (Johnson Div., UOP Inc., 1972, p. 91). This technique is only as accurate as the pressure gage being used.

Depth to water can also be determined with an acoustic or sonic type well probe. The sonic source and sensor are installed at the top of the well; a separate timing unit determines the time required for a sonic pulse to travel from the source to the water surface in the well and back. Depth to water is read directly from the display-timing unit. Accuracy is reportedly ±1 ft for a given temperature range in the well.

Pressure transducers measure the depth of submergence of the transducer and therefore can be used to determine indirectly the depth to water by lowering the transducer to a prescribed depth. The transducer is connected via electrical cable to a monitor at the surface from which the depth of submergence can be read directly as pressure. Accuracy varies as the range of the transducer varies but a standard resolution of 0.01% can be expected. Sensitivity therefore varies as 0.01% of the pressure range of a selected transducer. The sensitivity, as depth of water, decreases as the pressure range increases. Transducers are probably best suited to measuring changes in the water level with a semipermanent to permanent installation as would be expected in various test procedures. The transducers also can be tied into relay monitoring devices that can check rapidly the water level in several wells at the same time. Chart recorders can be obtained also.

Mechanically activated drum recorders may be used to obtain a continuous record of water level fluctuations. A battery clock or mechanical clock is used to drive a pen across a drum chart at a constant speed. Time increments per unit width of chart can be varied by changing the gear ratio on the clock drive mechanism. Water level changes actuate a counterweighted float that rotates the same drum chart. Gear ratios can be varied to suit an anticipated range of water level fluctuations as small as 0.001 ft (Davis, et al., 1975, p. 55). The depth to water is measured initially by chalked tape while subsequent readings can be taken directly from the chart. Water levels should be measured independently of the chart reading whenever the chart is changed.

Patton (1979) has discussed in considerable detail the importance of multiple measurements of hydraulic head with depth. The vertical distribution of pressure at a location can be illustrated graphically as shown on Fig. 31a, b, c. Fig. 31a illustrates a simple case involving a single hydrostratigraphic unit. A reference hydrostatic line is drawn first to illustrate the theoretical hydraulic pressure that would be measured at each observation port, assuming the ground water table or potentiometric surface is at the ground surface. This line would be at a 45° angle if the horizontal and vertical scales and units are equal. Pressure data measured along a vertical line will plot as a straight line parallel to the reference hydrostatic line if there is no vertical flow of water within the unit. Fig. 31a illustrates a simple case where the water table is encountered below the land surface but there is no vertical movement of water since the actual and reference hydrostatic lines are parallel.

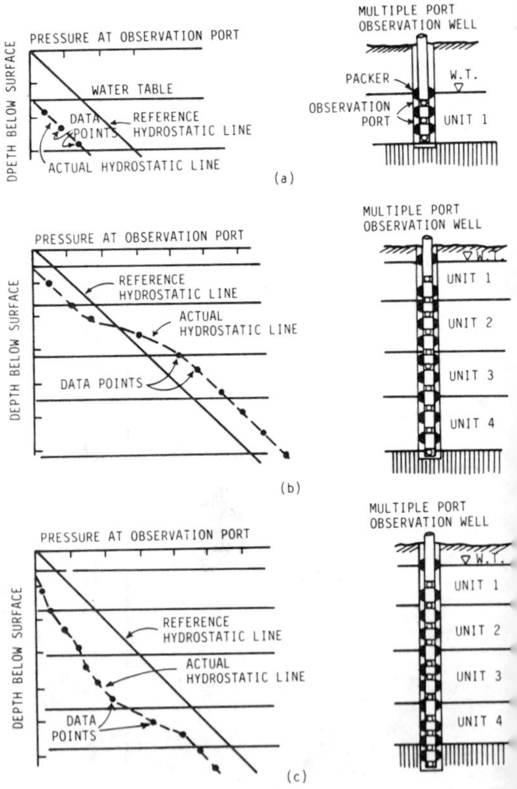

Fig. 31. Water level data evaluation (modified after Patton, 1979).

Fig. 31b illustrates a more complex situation involving several hydrostratigraphic units. Hydraulic head data from unit 1 indicate that there is no detectable vertical flow of water since the actual and reference hydrostatic lines are parallel. Unit 2 reflects a vertical flow of water from unit 3. The upward flow is indicated by a flatter slope on the measured hydrostatic line as compared to the reference hydrostatic line. A well penetrating unit 2 where the actual hydrostatic line is above the reference hydrostatic line can potentially flow at the ground surface since the hydraulic head is greater than the depth below the ground surface. There is no detectable vertical flow of water in or between units 3 and 4 since the actual hydrostatic line parallels the reference hydrostatic line.

Fig. 31c illustrates a complex case where there is downward movement of water from units 1 and 2 to unit 3. Downward movement of water is indicated by the actual hydrostatic line being steeper than the reference hydrostatic line. Unit 4 contributes upward flow of water to unit 3 as indicated by the flatter slope of the actual hydrostatic line versus the reference hydrostatic line. There is no apparent vertical flow below unit 4 since the actual and reference hydrostatic lines are parallel.

Hydraulically, these observations can be categorized into three fundamental concepts. No ver-

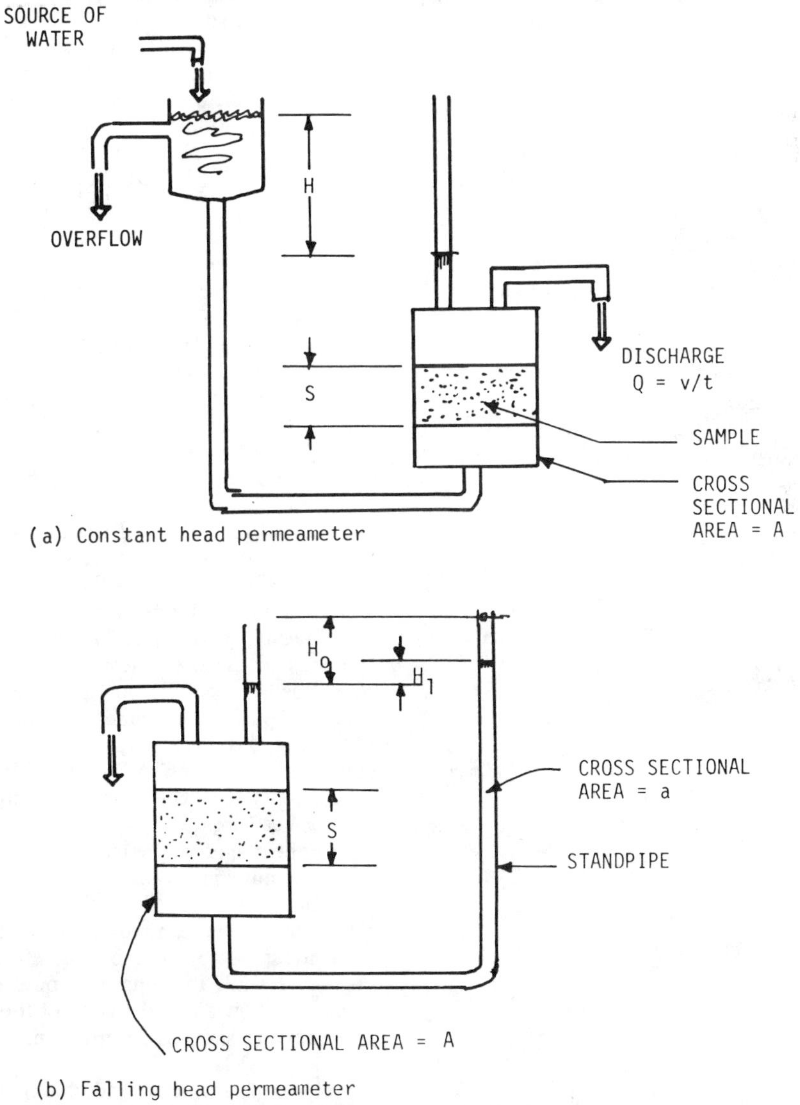

Fig. 32. Permeameters.

tical flow of ground water occurs if the actual hydrostatic line parallels the reference hydrostatic line. Upward flow occurs if the actual hydrostatic line is flatter than the reference hydrostratic line. Downward flow exists if the actual hydrostatic line is steeper than the reference hydrostatic line.

Two points of measurement are required to define a straight line relationship between hydraulic head and depth within any hydrostratigraphic unit. A third measuring point is required to determine if the relationship actually is a straight line while a fourth point is desirable for a minimum amount of redundancy. Numerous measuring points are required in a multi-unit system for defining the vertical distribution of hydraulic head and the vertical flow regime. This reasoning illustrates the justification and need for multiport piezometers.

Laboratory Methods: Hydraulic Conductivity: The hydraulic conductivity of a hydrostratigraphic unit can be determined in a laboratory from field samples or in the field by in situ testing. The laboratory techniques entail the use of either a constant head permeameter (Fig. 32a), a falling head permeameter (Fig. 32b), a radial percolation test (Fig. 33), or a consolidation test.

The constant head permeameter passes a constant flow of water through a sample of thickness

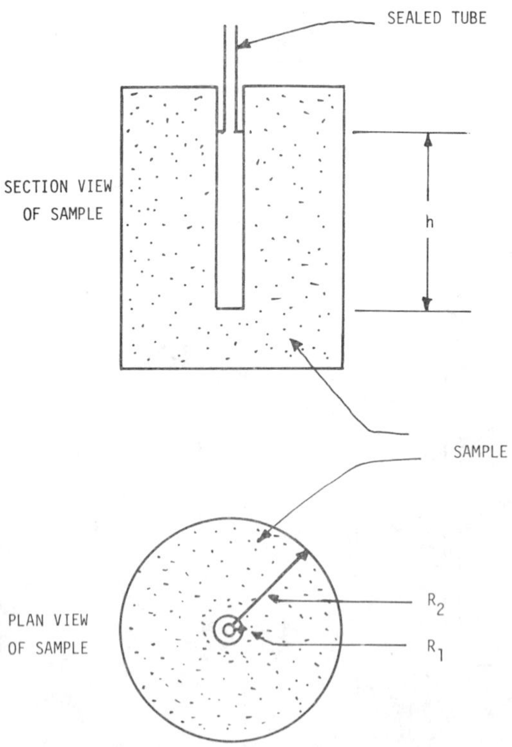

Fig. 33. Radial percolation test (modified after Jaeger, 1972, p. 60).

S and cross-sectional area A. The rate of flow and the head differential across the sample allow the calculation of hydraulic conductivity as follows:

$$K = \frac{QS}{AH} \tag{11}$$

where K is the hydraulic conductivity (L/T), Q is the steady volumetric discharge (L^3/T), S is the sample thickness (L), A is the cross-sectional area of sample (L^2), and H is the head differential across sample (L) (Freeze and Cherry, 1979), pp. 335–336).

A falling head permeameter allows the determination of hydraulic conductivity by measuring the rate of decline of the water level in the standpipe (Fig. 32b). A supply of water at a constant head is not used in this test but the head on the discharge side of the permeameter should be at a constant level. Hydraulic conductivity is calculated with the following equation:

$$K = \frac{aS}{AT} \ln\left(\frac{H_0}{H_1}\right) \tag{12}$$

where a is the cross-sectional area of the standpipe (L^2), S is the sample thickness (L), A is the cross-sectional area of sample (L^2), T is the time for decline of water level from H_0 to H_1 in standpipe (T), H_0 is the water level in standpipe when timed decline begins (L), and H_1 is the water level in standpipe at the end of time (T) (L) (Freeze and Cherry, 1979, p. 336).

Undisturbed samples should be used in either permeameter and the flow of water should be from the bottom to the top of the sample to minimize the effects of trapped air. A skin can develop on the surface of an unconsolidated sample that can adversely affect the determination of head differential across the sample. This can be alleviated in the constant head permeameter by measuring the head loss between two points in the interior of the sample separated by some new length s (Fig. 34) (Terzaghi and Peck, 1966, pp. 52–53). A direct substitution of s for S in Eq. 11 yields the hydraulic conductivity with this new arrangement for determining head loss.

Permeameter wall effects on the flow of water can be significant. Franzini (1956) determined that for granular samples the ratio of the permeameter diameter to the mean particle diameter should exceed about 40 to minimize the effect of the permeameter wall. Precautions should also be taken to prevent flow along the walls of the permeameter when testing samples from a consolidated formation.

The constant head and falling head permeameters can be utilized for measuring the hydraulic

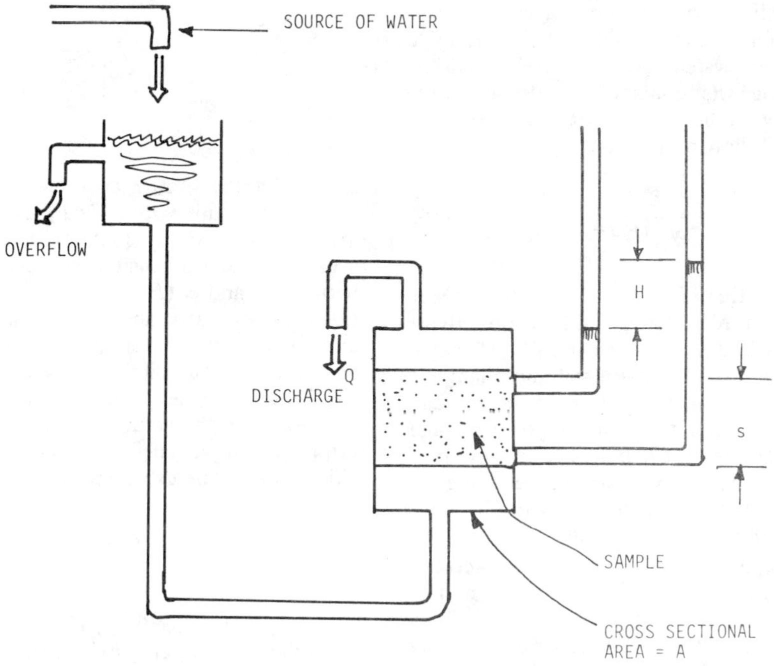

Fig. 34. Variation of constant head permeameter to alleviate skin effects on sample.

conductivity both parallel and perpendicular to bedding by altering the orientation of the sample in the permeameter. Selection of the appropriate test is based on the hydraulic conductivity of the sample. Klute (1965, p. 216) noted that the falling head permeameter is best suited to measuring a conductivity less than 0.01 cm/min, while a constant head permeameter can be used for conductivities greater than 0.01 cm/min as would be expected with sands and gravel.

A radial percolation test can be conducted on a rock sample in two different ways. A convergent flow test consists of placing water at a pressure greater than atmospheric around the exterior of the rock sample (Fig. 33). The tube, connected to the hole in the rock, is at atmospheric pressure, thereby creating a flow of water into the sample. This arrangement places the sample under compressive stress. The hydraulic conductivity can be determined as follows:

$$K = \frac{Q}{2\pi hp} \ln \frac{R_2}{R_1} \tag{13}$$

where K is the hydraulic conductivity (LT^{-1}), Q is the discharge of water percolating through the sample (L^3/T), h is the length of hole in sample (L), p is the pressure on exterior of sample above pressure in hole (L), R_1 is the radius of hole in sample (L), and R_2 is the radius of sample (L) (Jaeger, 1976, pp. 60–61).

A radial divergent flow test can be conducted by reversing the pressure differential. Water is introduced through the tube (Fig. 33) into the hole in the rock sample at a pressure greater than atmospheric. Flow occurs radially from the hole to the exterior of the sample that is at atmospheric pressure. This arrangement places the sample under tensile stress. Flow through the ends of the sample can be proven to be negligible (Jaeger, 1972, p. 61).

Tests conducted on the same sample under convergent and divergent flow conditions and with varying pressure differentials can yield valuable information. Rocks with closed pores have the same hydraulic conductivity for convergent and divergent flow so long as the tensile yield of the sample has not been reached (Jaeger, 1972, p. 61). Fissured rock samples exhibit a decreasing hydraulic conductivity with an increasing pressure for convergent flow, but this process is reversible until tensile failure occurs with divergent flow (Jaeger, 1972, p. 61). Appropriate procedures and equipment for conducting permeameter tests are described by Jaeger (Jaeger, 1972, pp. 56–64) and in the Special Technical Publication No. 417 of the American Society for Testing & Materials (1967).

The hydraulic conductivity of clayey materials can be determined from a consolidation test. A sample is loaded systematically, thereby causing transient drainage that controls the rate of consolidation. The hydraulic conductivity can be calculated from the following relationship:

$$T_v = \frac{K}{\gamma_\omega M_v} \frac{t}{H^2} \tag{14}$$

where T_v is the time factor for a given percent consolidation (T), K is the hydraulic conductivity (permeability) (L/T), γ_ω is the density of water (M/V), M_v is the coefficient of volume change, t is the time required to reach a given percentage consolidation (T), and H is the longest drainage path (L) (Cedergren, 1967, p. 42). The time rate of decline of sample thickness for each loading increment can be used to calculate hydraulic conductivity as demonstrated by Lambe (1951).

Laboratory Methods: Storage Characteristics: The storage coefficient of a confined hydrostratigraphic unit can be determined from potentiometric maps for that unit as outlined by Sahni and Seth (1979). This method requires prior knowledge of the transmissivity of that unit from some other means such as pump tests. Potentiometric maps are constructed for the unit in the area of interest for two different times separated by some arbitrary time interval, say four weeks. A profile is constructed along a line perpendicular to these contours where the least interpolation is required. The hydraulic gradient is determined at two points on this line separated by a distance that assures a significant difference in the hydraulic head between the points (Fig. 35). The storage coefficient can be determined from:

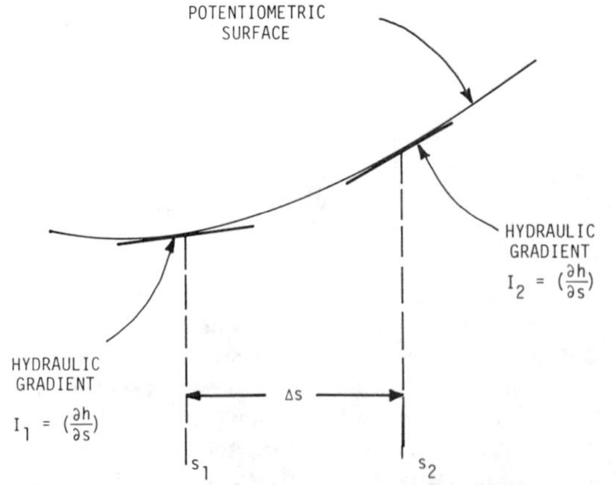

Fig. 35. Determining the storage coefficient from a changing potentiometric surface (modified after Sahni and Seth, 1979).

$$S = T \left| \frac{\frac{I_2 - I_1}{\Delta s}}{\frac{\Delta h}{\Delta t}} \right| \tag{15}$$

where S is the storage coefficient (dimensionless), T is the transmissivity (L^2/T), I_2 is the hydraulic gradient at s_2 (dimensionless), I_1 is the hydraulic gradient at s_1 (dimensionless), Δs is the distance between s_1 and s_2 (L), Δh is the change in potentiometric level at s_1 and s_2 that must be equal (L), and Δt is the time interval selected between sets of observations (T) (Sahni and Seth, 1979).

The specific yield of a unit can be approximated by measuring the volume of water drained from a saturated sample (such as a core) of that unit. Specific yield can be expressed as:

$$S_y = \frac{v_g}{V} \tag{16}$$

where S_y is the specific yield (dimensionless), v_g is the volume of water gravity drained from sample (L^3), and V is the total volume of sample (L^3) (Lohman, 1972, p. 6).

Field Methods: Hydraulic Conductivity and Storage Characteristics: Field methods for determining the hydraulic conductivity and storage characteristics of a given hydrostratigraphic unit are discussed in the following categories:

1) Single well tests.
2) Multiple well tests.
3) Fractured aquifer solution.
4) Vertical hydraulic conductivity.
5) Pump test recovery.
6) Boundary effects.
7) Geophysical techniques.
8) Computer techniques.

The available tests and procedures are summarized. The reader is referred to the corresponding reference cited for a more complete discussion on the application of the test and analysis of the data. Williams (1982) compares most of the methods in some detail.

Single well tests usually consist of either adding (slug test) or removing (bail test) a known volume of water from the well and measuring the change in water levels in the well with time. The same effect can be created by displacing a known volume of water in the well. The slug or bail tests create a water level response in the well that is identical for purposes of analysis except for the sign of the response. These types of tests are indicative of the hydraulic properties of the unit only in the immediate vicinity of the well due to the low volume of fluids involved. A single well can also be pumped at a constant rate while time-dependent water level measurements are made in the well. Single well pump tests require a separate method of analysis from the slug or bail tests; single well pump tests also require careful evaluation due to casing and borehole storage effects. Pump tests affect a greater volume of the aquifer due to the greater volume of fluids involved. Pump tests are more indicative of general hydrogeologic properties and usually vary less in a hydrostratigraphic unit than do point tests.

Single Well Test, Unconfined—The evaluation technique selected for either a slug or bail test is dependent on well geometry and whether the hydrostratigraphic unit being tested is confined or unconfined. An instantaneous change in the water level of a well in an unconfined unit is created by adding, removing, or displacing a known volume of water. Frequent measurements are made of the subsequent changing water level in the well with time. The hydraulic conductivity can be calculated by the equation developed by Bouwer and Rice (1976) and as described by Bouwer (1978, pp. 114–117):

$$K = \frac{r_c^2 \ln (R_e/r_w)}{2L_e} \frac{1}{t} \ln \left(\frac{y_o}{y_t}\right) \qquad (17)$$

where K is the hydraulic conductivity (L/T), r_c is the radius of the well section where the water level is changing (L), R_e is the effective radial distance over which the head difference is dissipated (L), r_w is the radial distance including thickness of gravel pack and developed zone outside the casing (L), L_e is the height of open section of well through which water moves (L), y_o is the static water level in the well prior to an instantaneous change in level (L), y_t is the water level in the well at some time (t) after the instantaneous change in water level (L), and t is the time since y_o when the instantaneous change in water level was induced (T) (Bouwer, 1978, pp. 114–117). The term $\ln R_e/r_w$ is determined from a plot of an analog analysis of well construction details and the position of the static water level. This plot and a more detailed discussion are presented in the references cited previously. This method is also applicable to units that receive water from an overlying confining layer. The well does not have to be fully penetrating for the application of this method but the solution for this test is based on the following assumptions:

1) There is negligible drawdown of the water level around the well.

2) The flow or movement of water above the water table can be ignored.

3) Well losses are negligible.

4) The hydrostratigraphic unit tested is homogeneous and isotropic (Bouwer and Rice, 1976).

An alternate method for evaluating a slug or bailer test in an unconfined unit was developed by Dagan (1978). The hydraulic conductivity is calculated from well completion details and measurements of the time and recovering water level in the well created by some rapid change in the water level. Hydraulic conductivity is determined from:

$$K = \frac{r_w^2}{2\overline{Q}L(t - t_o)} \ln \frac{(2L - y)/(2L - y_o)}{y/y_o} \qquad (18)$$

where K is the hydraulic conductivity (L/T), r_w is the radius of well (L), $\overline{Q}$ is determined from table of values of Ernst formula (dimensionless), L is the length of open section of well through which flow occurs (L), t is the time since change in static water level occurred (T), t_o is the time when the change in water level was induced (T), y is the distance from static water at time t_o to the water level at time t (L), and y_o is the change in water level in the well at time t_o (Dagan, 1978). A table of values for $\overline{Q}$ and a more detailed discussion of this method are presented by Dagan (1978). This method does not permit adjustment or compensation for gravel packed wells or wells with a significant developed zone extending beyond the immediate well radius. The well can be partially or fully penetrating but the perforated portion of the well length should be at least 50 times the well radius (Dagan, 1978). The Laplace equation was used in the development of this technique that requires that the medium be considered homogeneous and isotropic. It also requires that the potential at the water table be constant (Dagan, 1978).

Hvorslev (1951) presented a thorough discussion of in situ tests for determining hydraulic conductivity with various configurations of test well completions. The methods presented are applicable to confined units of infinite thickness or conditions such that the inflow or outflow of water in the well does not cause any appreciable change in the ground water level (Hvorslev, 1951, p. 30). Water injected in a well at a constant head allows the determination of hydraulic conductivity by the following relationship:

$$K = \frac{q}{FH_c} \qquad (19)$$

where K is the hydraulic conductivity (L/T), q is the rate of flow (L^3/T), F is the shape factor presented by Hvorslev (1951), and H_c is the constant head at which water is injected into the well (L) (Hvorslev, 1951, p. 43).

The hydraulic conductivity can also be determined from tests where the water level in the well fluctuates due to the addition or removal of a volume of water. The hydraulic conductivity can be calculated with the following equation:

$$K = \frac{A}{F(t_2 - t_1)} \ln \frac{H_1}{H_2} \qquad (20)$$

where K is the hydraulic conductivity (L/T), A is the cross-sectional area of standpipe (L^2), F is the shape factor presented by Hvorslev (1951), t_1 is some time after the change induced in the water level of the well (T), t_2 is some time after t_1 measured from the induced change in the water level of the well (T), H_1 is the difference in water level in the well between static conditions and time t_1 (L), and H_2 is the difference in water level in the well between static conditions and time t_2 (L) (Hvorslev, 1951).

It is recommended that a semilogarithmic plot be made of water level ratios (H_1/H_2) and time to verify that the data plots are a straight line through the origin of the diagram. This verifies that the assumptions are fulfilled (Hvorslev, 1951, p. 13). The hydraulic conductivity can also be determined from a variation of Eq. 20 if the time lag has been determined from the semilogarithmic plot. These procedures are also presented by Cedergren (1967, pp. 81–86).

Single Well Test, Confined—Cooper (1963) presented a method of analysis applicable to confined units where the well is open to the full thickness of a hydrostratigraphic unit. Again, an instantaneous change in the water level in the well is required as are frequent, timed water level measurements after the instantaneous change. The data are plotted on semilogarithmic graph paper as a ratio of the change in water level to the maximum change in water level versus time. This plot is compared to a set of type curves that can be prepared from tables presented by Cooper et al. (1967) or as can be found in Lohman (1972). An extension of the type curves was prepared by Papadopulos et al. (1973) for units exhibiting very low storage coefficients. A value for the time of measurement is obtained from the data-type curve match and inserted in:

$$T = \frac{1.0\, r_c^2}{t} \qquad (21)$$

where T is the transmissivity (L^2/T), r_c is the casing radius over which water level fluctuates (L), and t is the time of measurement (T) (Lohman, 1972, pp. 27–29).

The storage coefficient can also be determined by this method with:

$$S = \frac{r_c^2}{r_s^2}\alpha \qquad (22)$$

where S is the storage coefficient (dimensionless), r_c is the casing radius over which water level fluctuates (L), r_s is the radius of open hole or well screen (L), and α is the coefficient determined from type curve match (dimensionless) (Lohman, 1972, pp. 27–29). According to Cooper et al. (1967) the reliability of this method of determining the storage coefficient is questionable.

Lohman (1972, pp. 29–30) illustrates an alternative method for calculating transmissivity from a well that has been repeatedly bailed; the method of analysis was originally proposed by Skibitzke. The time intervals between bail cycles and the volume of water removed during the cycle are noted. This test should give satisfactory estimates for transmissivity in a confined unit where the water level is shallow enough to permit a short time interval between bail cycles. For large values of time the transmissivity can be calculated from:

$$T = \frac{1}{4\pi s'}\left(\frac{V_1}{t_1} + \frac{V_2}{t_2} + \frac{V_3}{t_3} + \ldots + \frac{V_n}{t_n}\right) \qquad (23)$$

where T is the transmissivity (L^2/T), s' is the residual drawdown or difference between the prevailing water level and the water level in the well at some time t after bailing commenced (L), $V_{1,2,\ldots,n}$ is the volume of water removed at each bail cycle (L^3), and $t_{1,2,\ldots,n}$ is the elapsed time between bail cycle and measurement of s' (Lohman, 1972, pp. 29–30).

This equation can be simplified if nearly equal volumes of water are removed with each cycle. The equation becomes:

$$T = \frac{V}{4\pi s'}\left(\frac{1}{t_1} + \frac{1}{t_2} + \frac{1}{t_3} + \ldots + \frac{1}{t_n}\right) \qquad (24)$$

where all the variables are as previously outlined for Eq. 23 (Lohman, 1972, pp. 29–30).

Hydrostratigraphic units with a very low hydraulic conductivity require special techniques for in situ testing by the slug type test. Bredehoeft (1980) outlined two techniques that can be used to reduce the excessively long monitoring periods required for the determination of transmissivity. One technique requires the installation of a small diameter monitoring tube (standpipe) in the well as illustrated in Fig. 36. A unit decline in the water level in the standpipe represents a much smaller volume exchange of water into or out of the hydrostratigraphic unit of interest than if the water level was declining within the original well casing as utilized in the previously described slug test procedures. The time rate decline of the water level is analyzed by Cooper et al. (1967) and Papadopulos et al. (1973) and was discussed earlier. This method

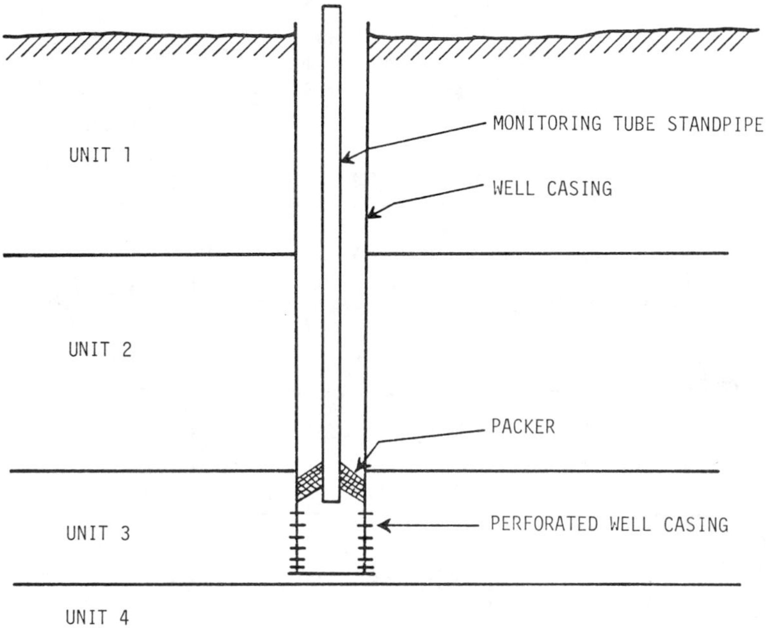

Fig. 36. Single well schematic for slug-type test of low hydraulic conductivity units.

also has its limitations; the diameter of the standpipe can become too small for measuring water level declines if the hydraulic conductivity of the unit is extremely low.

A second alternative exists for these cases. The well is modified as illustrated in Fig. 37. The well is filled with water and after observing the water level for a period of time, the system is pressurized by injecting more water with a high pressure pump. The system is shut in and the pressurization allowed to decay. This procedure assumes that there is an insignificant drop in the water level in the well after the initial filling or that the decline in the water level can be extrapolated through the duration of the test without significant errors. Observed values of the ratio of head decline to maximum induced head change are plotted vs. time on semilogarithmic graph paper for type curve matching. This technique is described by Cooper et al. (1967) and Papadopulos et al. (1973). Transmissivity can be calculated by substituting the appropriate type curve data and water properties into:

$$T = \frac{1.0\ V_\omega\ C_\omega\ \rho_\omega\ g/\pi}{t} \qquad (25)$$

where T is the transmissivity (L^2/T), V_ω is the volume of water within the pressurized section of the well (L^3), C_ω is the compressibility of water (LT/M), ρ_ω is the density of water (M/L^3), g is the acceleration due to gravity (L/T^2), and t is the time of measurement since pressurization (T) (Bredehoeft and Papadopulos, 1980).

The storage coefficient is calculated from:

$$S = \frac{V_\omega\ C_\omega\ \rho_\omega\ g/\pi}{r_s^{\ 2}} \qquad (26)$$

where S is the storage coefficient (dimensionless), V_ω is the volume of water within the pressurized section of the well (L^3), C_ω is the compressibility of water (LT^2/M), ρ_ω is the density of water (M/L^3), g is the acceleration due to gravity (L/T^2), α is the coefficient determined from type curve match (dimensionless), and r_s is the radius of open hole or well screen (L) (Bredehoeft and Papadopulos, 1980). Again, the value calculated for the storage coefficient by this method is of questionable reliability when the match curve value for α is 0.1.

A different approach is required for data-type curve matches for α greater than 0.1 since the similarity of the type curves is much greater and hence the possible error in the determined transmissivity is as great as that for the storage coefficient. It is possible only to determine the product of transmissivity and storage coefficient in this range with:

$$TS = \frac{\alpha\ B(V_\omega\ C_\omega\ \rho_\omega\ g)^2}{\pi^2 r_s^{\ 2} t} \qquad (27)$$

where T is the transmissivity (L^2/T), S is the storage coefficient (dimensionless), αB is the product

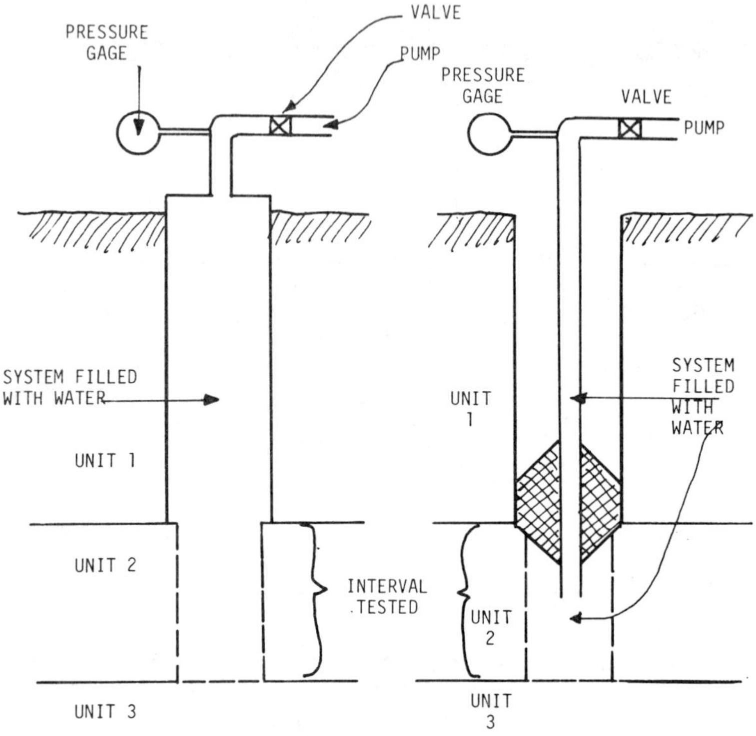

Fig. 37. Single well schematic for slug-type test of extremely low hydraulic conductivity units (after Bredehoeft and Papadopulos, 1980).

parameter determined from type curve match (dimensionless), V_ω is the volume of water within the pressurized section of the well (L^3), C_ω is the compressibility of water (LT^2/M), ρ_ω is the density of water (M/L^3), g is the acceleration due to gravity (L/T^2), r_s is the radius of open hole or well screen (L), and t is the time of measurement since pressurization (T) (Bredehoeft and Papadopulos, 1980). Two critical assumptions that must be met for this method of testing are:

1) Flow in the test interval is primarily radial.

2) The hydraulic properties of the unit tested remain constant throughout the test.

Care must be exercised to prevent overpressurizing the unit since that can cause fractures to dilate or cause an alteration of the local hydraulic conductivity. A complete discussion of this method is presented by Bredehoeft and Papadopulos (1980).

The petroleum industry variation of these tests is referred to as the drill-stem test. Hackbarth (1978) noted that drill-stem tests become mechanically reliable and economical where single measurements of the hydraulic head and hydraulic conductivity are desired at depths in excess of 1000 ft (305m). A water sample can also be obtained from the zone tested after removing the drill-stem test equipment that is commonly removed after each test. Recording charts in the pressure sensing devices are checked for possible defects in their mechanical operation prior to their reinstallation in the drill-stem and placement in the well. Typical drill-stem equipment is shown in Fig. 38 for a bottom hole test. Zone testing above the bottom of the well would require an additional packer on the bottom of the drill stem. The drill-stem test is not applicable where the hydrostratigraphic unit to be tested has a relatively high hydraulic conductivity and a low pressure head (low water level).

The pressure responses shown on a chart from a typical drill-stem test are shown in Fig. 38. The initial increase in pressure is due to the lowering of the drill-stem test equipment in the well as illustrated by A-B in Fig. 38. The packers are expanded at the desired depth and the system allowed to stabilize prior to opening the test valve. Fluid flows into the air filled drill-stem (C-D) and the valve is closed. The pressure response after shut-in is analyzed for hydraulic conductivity. Additional tests can be conducted by again opening the valve and then closing it (E-F-G). The shut-in pressure response can be evaluated for each test. A water sample is obtained by leaving the valve closed and

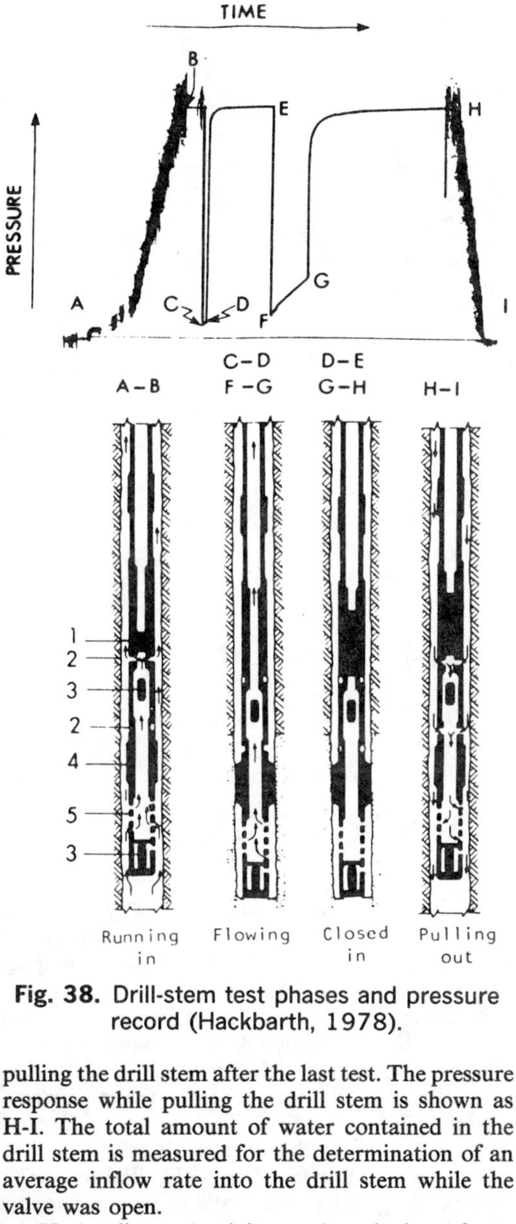

Fig. 38. Drill-stem test phases and pressure record (Hackbarth, 1978).

pulling the drill stem after the last test. The pressure response while pulling the drill stem is shown as H-I. The total amount of water contained in the drill stem is measured for the determination of an average inflow rate into the drill stem while the valve was open.

Hydraulic conductivity can be calculated from:

$$K = \frac{2.303q}{4\pi h(P_o - P_\omega)} \log \frac{t + \Delta t}{\Delta t} \qquad (28)$$

where K is the hydraulic conductivity (L/T), q is the rate of production during flow periods (L^3/T), h is the saturated thickness of unit tested (L), P_o is the reservoir pressure which can be converted to hydraulic head (L), P_ω is the shut-in pressure at time t that can be converted to hydraulic head (L), t is the total flowing time (T), and Δt is the time since shut-in started (T) (after Hackbarth, 1978). Data are plotted as shut-in pressure converted to feet of head vs. the logarithm of $t + \Delta t/t$. The use of base 10 logarithm simplifies the plot that should appear as a straight line. Hydraulic conductivity can be calculated by inserting the pressure change (converted to hydraulic head) per log cycle change in $t + \Delta t/t$ into Eq. 28. This solution assumes that the hydrostratigraphic unit tested is confined, homogeneous, isotropic, and infinite in areal extent. The well can be assumed to be infinitesimally small in diameter and water is removed instantaneously from storage. The rate of production is assumed to be constant for all flow periods.

Single wells may be pumped at a constant rate while the water level is monitored in the same well. These data can be analyzed by techniques summarized in the following section on Multiple Well Tests but caution should be exercised due particularly to the effects of storage and well losses associated with the pumping well. Storage and fracture interception effects are discussed in the section on Fracture Solutions.

Single Well Test, Borehole Dilution—This test requires the injection and monitoring of a tracer in a single well. A selected portion of a borehole or well is isolated and injected with a known concentration of tracer that is kept thoroughly mixed in the borehole throughout the test. The reduction in concentration of the tracer is noted as the natural ground water movement removes the tracer from the well. An apparent ground water velocity can be calculated based on the well dimensions. This apparent velocity can be converted to an average ground water velocity with an adjustment factor that is based on the geometry of the well screen and gravel pack, if present (Freeze and Cherry, 1979, pp. 429–430).

Multiple Well Tests—Pump tests consist of discharging water at a constant rate from a well that is open within the hydrostratigraphic unit of interest. Weeks (1978) discusses the state of the art for these types of tests. In Fig. 39, the decline in water level within unit 2 is measured in the pumping well, and in wells A and B. This areal decline in the water level is referred to as the cone of depression. It is depicted in Fig. 39 as the water level at time t_1 and later at time t_2. The decline in water levels and the expansion of the areal extent of the cone of depression theoretically will continue indefinitely if unit 2 is infinite in areal extent and fully artesian without leakage from unit 1.

The decline in water levels from the prepumping water level is measured and recorded with respect to the time since discharge began. The data can be plotted on log-log paper or semilog paper depending upon the method of analysis chosen (Fig. 40). Data plots reflect the hydrogeologic conditions

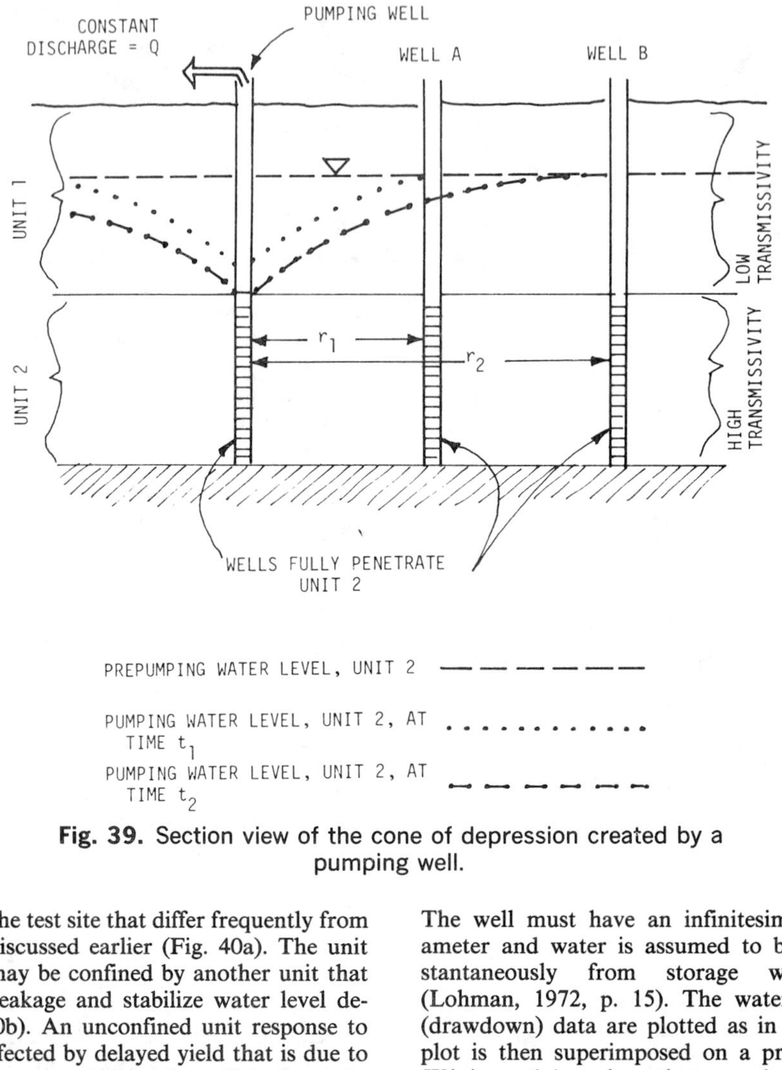

Fig. 39. Section view of the cone of depression created by a pumping well.

prevailing at the test site that differ frequently from the ideal as discussed earlier (Fig. 40a). The unit being tested may be confined by another unit that will provide leakage and stabilize water level declines (Fig. 40b). An unconfined unit response to pumping is affected by delayed yield that is due to vertical flow components near the well during early pumping. Delayed yield tends to stabilize water level declines until the delayed yield begins to diminish, at which time water level declines increase. The response to pumpage is illustrated by Fig. 40c. A hydrostratigraphic unit may be bounded by one or more boundaries that may be impermeable (very low hydraulic conductivity) and act as a barrier to the movement of ground water. Alternatively the unit may be bounded by a source of water at a constant head that will stabilize water level declines. Both cases are illustrated in Fig. 40d.

Multiple Well Tests, Nonleaky Artesian—The Theis solution is the traditional method of analysis for the non-steady state behavior of a well in a confined unit during pumping. This solution is applicable to a fully penetrating well in a unit that is homogeneous, isotropic, and infinite in areal extent. The well must have an infinitesimally small diameter and water is assumed to be removed instantaneously from storage with pumping (Lohman, 1972, p. 15). The water level decline (drawdown) data are plotted as in Fig. 40a. This plot is then superimposed on a prepared plot of $W(u)$ vs. $1/u$ values that may be found in numerous sources (Bouwer, 1978; Davis and DeWiest, 1966, 1967; Domenico, 1972; Freeze and Cherry, 1979; Lohman, 1972; Johnson Div., UOP, 1972; US Bureau of Reclamation, 1977; Walton, 1970; Weeks, 1978). The type curve is matched to the data plot while keeping both axes parallel. Values of $W(u)$, $1/u$, t, and s are read from a match point on the plots and substituted into:

$$T = \frac{Q}{4\pi s} W(u) \tag{29}$$

where T is the transmissivity (L^2/T), Q is the uniform well discharge (L^3/T), s is the drawdown in the observation well at a specific time (L), and $W(u)$ is the well function (dimensionless) (US Bu-

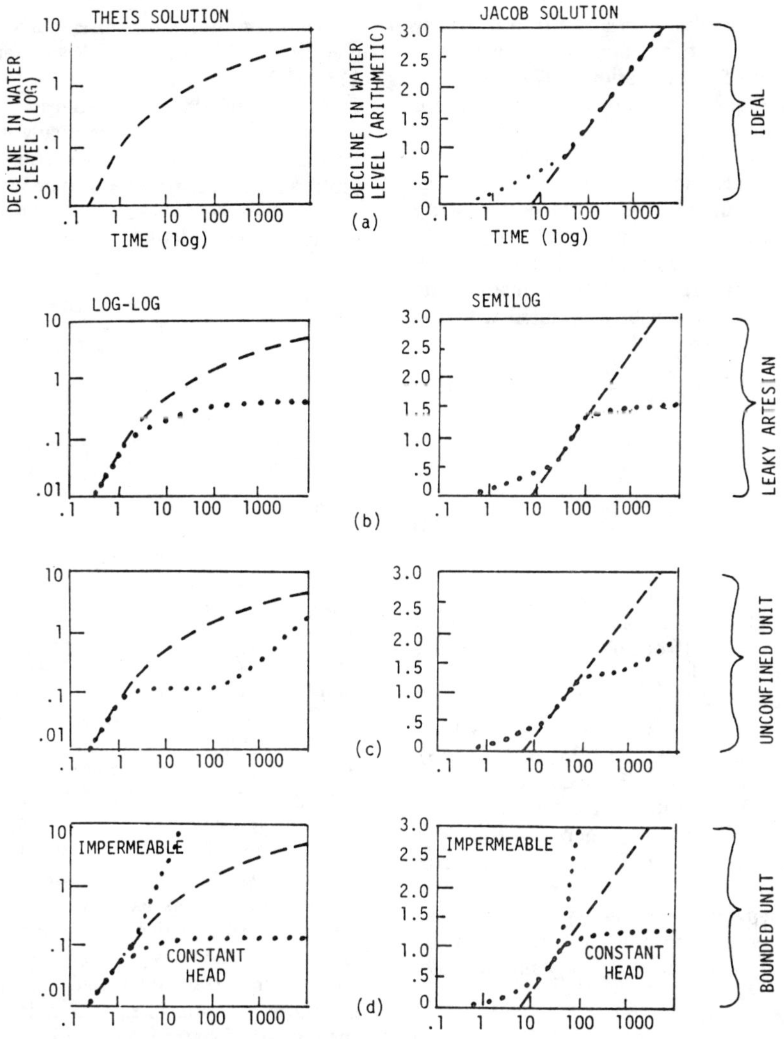

Fig. 40. Log-log and semilog plots of pump test data under various hydrogeologic conditions.

reau of Reclamation, 1977). The storage coefficient can be obtained by substituting into:

$$S = \frac{4Tut}{r^2} \tag{30}$$

where S is the storage coefficient (dimensionless), T is the transmissivity (L^2/T), u is the well function (dimensionless), t is the time since pumping started (T), and r is the radius to the observation well (L) (US Bureau of Reclamation, 1977). Data may be plotted also as s vs. t/r^2 where drawdown data from multiple observation wells exist. These data are matched to a type curve of $W(u)$ vs. $1/u$ and match point values are substituted in Eqs. 29 and 30 (Lohman, 1972, p. 18). A more complete discussion is presented by Lohman (1972) and other references (US Bureau of Reclamation, 1977).

An approximate solution to the previously described Theis solution is known as the straight line method utilizing semilog paper (Fig. 40a-d). In order for this approach to be valid u must be equal to or less than 0.01 (Lohman, 1972, p. 19) where u is defined as

$$u = \frac{r^2S}{4Tt} \tag{31}$$

where u is the well function (dimensionless), r is the radius to observation well (L), S is the storage coefficient (dimensionless), T is the transmissivity (L^2/T), and t is the time since pumping started (T) (US Bureau of Reclamation, 1977, pp. 101–112). The time required for the data to be valid for this solution increases as the distance between the pumping well and the observation well increases. The time factor also increases linearly with the storage coefficient and inversely with transmissivity (Lohman, 1972, p. 22).

Drawdown data may be plotted vs. time (Fig. 41) for each well where $u \leq 0.01$. Transmissivity and storage coefficient can be calculated with:

$$T = \frac{2.303Q}{4\pi \Delta S} \tag{32}$$

and

$$S = \frac{2.25Tt_o}{r^2} \tag{33}$$

where T is the transmissivity (L^2/T), Q is the uniform discharge (L^3/T), Δs is the drawdown over one log cycle (L) (from plot), S is the storage coefficient (dimensionless), t_o is the time at zero drawdown intercept (T) (from plot), and r is the distance from pumping well to observation well (L) (US Bureau of Reclamation, 1977, pp. 101–113). Time drawdown data are influenced by the interception of boundaries as illustrated by Fig. 40d. A barrier boundary theoretically increases the slope of the semilog data plot by a factor of two whereas a recharge boundary reduces the slope by a factor of two. A linear recharge boundary may approach but never attain a condition of zero additional drawdown with time but delayed yield or leaky

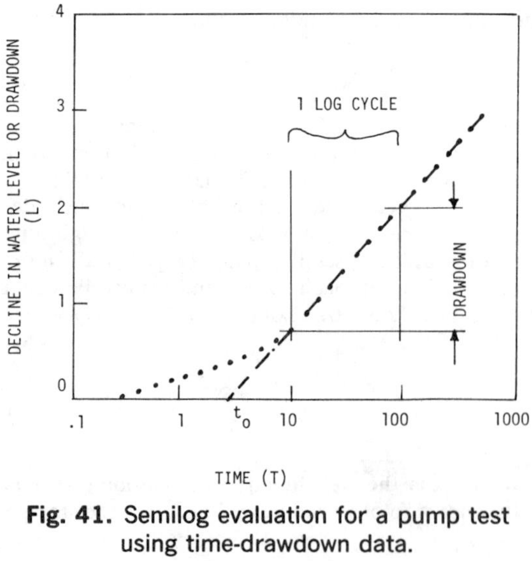

Fig. 41. Semilog evaluation for a pump test using time-drawdown data.

artesian conditions may create a stable drawdown condition (US Bureau of Reclamation, 1977, pp. 113–114).

A distance drawdown solution can be obtained where drawdown data are available for the same time from at least three observation wells. Data are plotted as drawdown vs. distance from the pumping well as shown in Fig. 42. Transmissivity and storage coefficient are calculated from:

$$T = \frac{2.303Q}{2\pi \Delta s} \tag{34}$$

and

$$S = \frac{2.25\,Tt}{r^2_o} \tag{35}$$

where T is the transmissivity (L^2/T), Q is the uniform discharge (L^3/T), s is the drawdown over one log cycle (L) (from plot), t is the time since pumping began (T), and r_o is the zero drawdown intercept (L) (from plot) (US Bureau of Reclamation, 1977, pp. 101–113).

The methods of analysis discussed in the following Leaky Artesian Solutions section basically are applicable to the nonleaky artesian analysis if leakage from the confining layer is assumed to equal or approach zero. Analysis with the leaky artesian solutions indicates whether this assumption is valid.

Multiple Well Tests, Leaky Artesian—Leaky artesian conditions exist when the confining unit has sufficient vertical hydraulic conductivity to transmit water. Few, if any, confining units exist that will not transmit water to an adjacent hydrostratigraphic unit when that unit is pumped. This movement of water from another unit further complicates the mathematical analysis of pump test data, but the analysis of leaky artesian conditions is simplified by assuming that flow in the confining unit is vertical while flow is horizontal in the higher hydraulic conductivity unit being pumped.

The modified method developed by Hantush (Hantush, 1960) assumes that the unit being pumped is homogeneous, isotropic, infinite in areal extent, uniform in thickness, and that water is released instantaneously from storage with a decline in water level. This modified method is very versatile as described by Weeks (1978) and by Lohman (1972, pp. 32–34). As leakage from the confining layer approaches zero, the general equation for a leaky aquifer becomes the Theis nonleaky artesian solution. Hantush's (1960) early solutions to the leaky artesian case accounted for the leakage from the confining unit but not for the release of water from storage in the confining unit. The modified method (Hantush, 1960) does account for storage

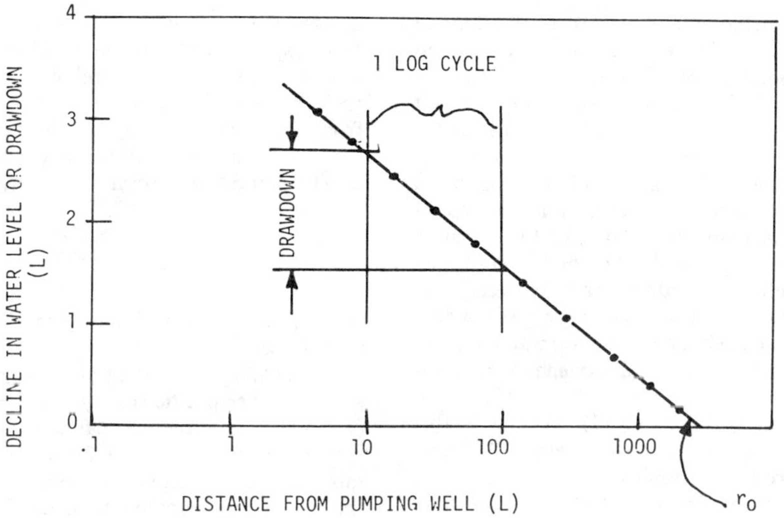

Fig. 42. Semilog evaluation for a pump test using distance-drawdown data.

in the confining unit that is a more realistic approach. Cooper (1963, p. c55) noted that the water yielded by a leaky artesian aquitard is largely derived from storage since the hydraulic gradient across the aquitard lags behind the decline in water level in the pumped unit.

Pump test data from a constant discharge test are plotted on log-log paper as water level decline (drawdown) vs. time since pumping began. The plot is superimposed on a prepared type curve of l/u vs. $H(u,B)$ such as Plate 4 of Lohman (1972). Match point data for four parameters are inserted into:

$$T = \frac{Q}{4\pi s} H(u,B) \tag{36}$$

and

$$S = \frac{4Ttu}{r^2} \tag{37}$$

where T is the transmissivity (L^2/T), Q is the constant discharge (L^3/T), s is the drawdown (L), $H(u,B)$ is the H function of u, B (dimensionless), S is the storage coefficient (dimensionless), t is the time since pumping began (T), u is the well function (dimensionless), and r is the radius to the observation well (L) (Lohman, 1972, pp. 32–34). A more complete explanation and an example are presented in Lohman (1972, pp. 32–34). Tables for preparing type curves are included in Hantush (1960).

Hantush (1959) developed an equation for determining the transmissivity and storage coefficient of a leaky artesian unit based on the free flow discharge of a well. The vertical hydraulic conductivity of the confining unit may also be determined with this procedure. Radial flow is assumed in an isotropic leaky artesian unit with fully penetrating wells. Water is not released from storage in the confining unit. The water level drops instantaneously with the initiation of well flow and is maintained at that level for the duration of the test (Walton, 1970, p. 148). Discharge is plotted on log-log paper vs. time since discharge began and this plot is superimposed on type curves such as those presented by Lohman (1972, Plate 5). Transmissivity and storage coefficient are calculated from match point values utilizing:

$$T = \frac{Q}{2\pi s_\omega G(\alpha, r_{\omega/B})} \tag{38}$$

and

$$S = \frac{Tt}{r^2_\omega \alpha} \tag{39}$$

where T is the transmissivity (L^2/T), Q is the well discharge at time t (L^3/T), s_ω is the drawdown in discharging well (L), $G(\alpha, r_{\omega/B})$ is the well function (dimensionless), S is the storage coefficient (dimensionless), t is the time since discharge began (T), r_ω is the radius of discharging well (L), and α is the well function (dimensionless) (Lohman, 1972, pp. vi–vii, 34).

Lohman (1972, p. 34) presents a more detailed discussion of the use of this method of analysis. Hantush (1959) presents tabular data for the construction of type curves.

A significant modification to these methods of analysis is presented by Neuman and Witherspoon (1972). This method requires the installation of observation wells in the hydrostratigraphic unit being pumped as well as the confining units above and/or below the pumped unit. The ratio of the drawdown in the confining unit to that measured in the pumping unit at the same time and radial distance from the pumping well is used to determine the hydrogeologic characteristics of the confining unit. This method (Neuman and Witherspoon, 1972) is applicable to leaky, multiple units where the pumped unit is either confined or unconfined. The confining units can be heterogeneous and anisotropic.

Neuman and Witherspoon (1972) refer to this analysis technique as the "ratio method." Drawdown values are obtained in the unit being pumped as well as in the confining unit at the same time and radial distance from the pumping well (Fig. 43). Drawdown data from observation wells within a few hundred feet of the pumping well are adequate for most cases. A dimensionless time value for the pumped unit is calculated from:

$$t_D = \frac{Tt}{Sr^2} \tag{40}$$

where t_D is the dimensionless time for the pumped unit, T is the transmissivity of the pumped unit (L^2/T), t is the pumping time (T), S is the storage coefficient of the pumping unit (dimensionless), and r is the radial distance from pumping to observation well (L). The drawdown ratio and the dimensionless time (t_D) are located in Fig. 3 of Neuman and Witherspoon (1972) and a value for the dimensionless time (t_D) of the confining unit is obtained.

The vertical hydraulic conductivity of the confining unit can be determined once the specific storage for the confining unit is known. Neuman and Witherspoon (1972) recommend determining specific storage by using borehole extensometers although approximate values can also be obtained from standard consolidation tests performed on core samples. The vertical hydraulic conductivity can be determined from:

$$t'_D = \frac{K't}{S'_s Z^2} \tag{41}$$

where t'_D is the dimensionless time for the confining unit, K' is the hydraulic conductivity in a vertical direction (L/T), t is the time since pumping began (T), S_s is the specific storage for the confining unit (l/L), and Z is the vertical coordinate (L). Neuman and Witherspoon (1972) suggest that the transmissivity and storage coefficient be determined for the unit being pumped with data obtained from observation wells close to the pumping well or from the pumping well itself if the effective radius of the well is known. Wells in hard rock formations would suffice. The Theis method of analysis can be applied to such data with only a small error being introduced.

Multiple Well Tests, Unconfined—Few wells can be considered fully penetrating, i.e., open to the unit over its entire saturated thickness (Fig. 44a). Most wells in an unconfined unit are open to

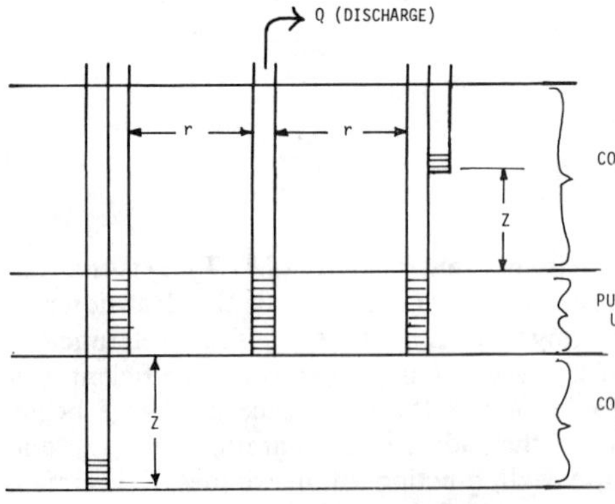

Fig. 43. Observation well schematic for determining vertical hydraulic conductivity of a confining unit (after Neuman and Witherspoon, 1972).

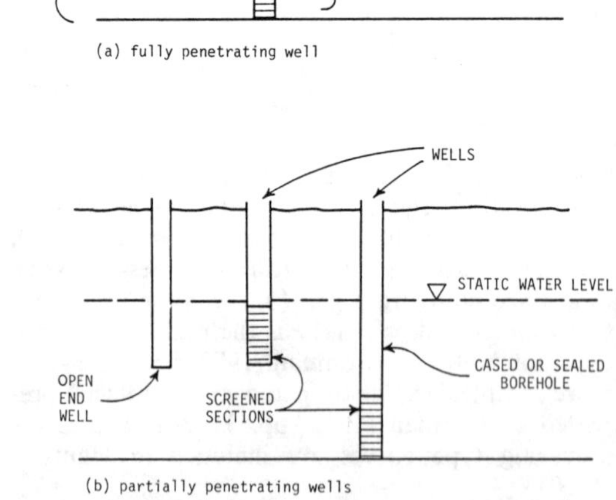

Fig. 44. Examples of fully and partially penetrating wells.

only a portion of the saturated thickness (Fig. 44b). Pumping wells frequently are open only to the lower third of the unit; that causes convergent flow rather than true radial flow. Under this condition water levels in observation wells open at the same position in the unit as the pumping well are lower than the average water level in the unit. Water levels higher than average for the hydrostratigraphic unit are measured in those observation wells open at a different position in the unit than the pumping well. These effects can be minimized by placing observation wells at least 1.5 times the saturated unit thickness away from the pumping well (US Bureau of Reclamation, 1977, p. 94). Alternatively observation wells can be placed in the upper and lower 15% of the saturated unit thickness and their water levels averaged for computational purposes (US Bureau of Reclamation, 1977, p. 94).

Most solutions to equations describing flow in artesian aquifers assume that the flow to the pumping well is radial. Unconfined units can be evaluated using the Theis solution for artesian aquifers if the drawdown is small in comparison to the overall saturated thickness of the unit. It has been suggested that this approach induces little error where the drawdown is less than 10% of the initial saturated thickness. A correction factor should be applied to drawdowns that exceed 10% but alternate solutions should be sought when drawdowns exceed 25% of the initial saturated thickness (US Bureau of Reclamation, 1977, p. 95). Drawdown data that fall in the 10 to 25% range can be adjusted with:

$$s' = s - \frac{s^2}{2M} \qquad (42)$$

where s' is the corrected drawdown (L), s is the measured drawdown (L), and M is the saturated thickness of the unit prior to pumping (L) (Walton, 1962, 1967, p. 7). Corrected data are plotted and evaluated as described previously with the exception that the storage coefficient now represents the specific yield of the hydrostratigraphic unit. Neuman (1975) found that this correction factor is not applicable to early or intermediate time drawdown where delayed gravity response occurs.

Stallman (1965) developed type curves for evaluating the early drawdown data associated with an unconfined aquifer pump test. These curves account for the variable effects of partial penetration in an aquifer before steady state drawdown conditions are approached. Lohman (1972, Plates 6 and 7) presents these type curves in a format suitable for matching data plots on log-log paper. Data can be plotted as drawdown vs. time (s vs. t), time divided by the radius squared (s vs. t/r^2), or the reciprocal of the radius squared for a constant time (s vs. $1/r^2$). Constant discharge is assumed. The data plot is superimposed and match point values selected such that sT/Q and Tt/r^2S are equal to one. Transmissivity and storage coefficient are calculated from:

$$T = 1.0\,\frac{Q}{s} \qquad (43)$$

and

$$S = \frac{Tt}{1.0\,r^2} \qquad (44)$$

where T is the transmissivity (L^2/T), Q is the constant discharge (L^3/T), s is the drawdown (L), t is the time since pumping began (T), and r is the radial distance (L) (Lohman, 1972, p. 35). The ratio of radial to vertical hydraulic conductivity can be determined from this type curve procedure using:

$$\psi = \frac{r}{b}\sqrt{\frac{K_z}{K_r}} \qquad (45)$$

where ψ is the type curve match value (dimensionless), r is the radius to observation well from pumped well (L), b is the thickness of saturated unit (L), K_z is the vertical hydraulic conductivity (L/T), and K_r is the radial hydraulic conductivity (L/T) (Lohman, 1972, pp. vi–vii, 35–38). A more detailed discussion is presented in Lohman (1972, pp. 35–38).

If delayed yield occurs during an aquifer test, an alternative procedure for analysis of data is required. Pumping from an unconfined unit creates a zone of dewatered material adjacent to the pumping well that expands as pumping continues. The zone of dewatering is not produced by instantaneous removal of all water within the zone. The initial response to pumping reflects the instantaneous release of water from storage by the compaction of the unit and the expansion of the water, but later responses, as illustrated in Fig. 40c, are affected significantly by delayed yield. The effects of delayed yield dissipate with time and the drawdown response approaches the nonleaky artesian response curve derived by Theis for later times as illustrated in Fig. 45.

The delayed yield response was described mathematically and tables of solutions presented by Boulton (1963). These solutions are presented in a type curve format suitable for data superimposition in (Lohman, 1972, Plate 8). This method of analysis assumes that the aquifer is homogeneous, isotropic, constant thickness, infinite in areal extent, the wells fully penetrating, the water in the unit is unconfined, and the drawdown is very small in compar-

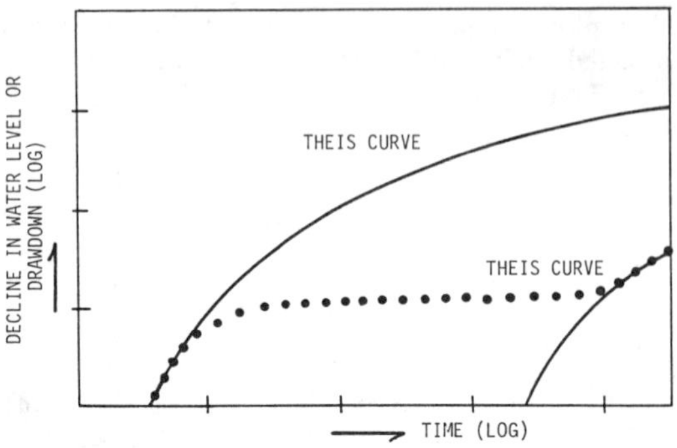

Fig. 45. Illustration of delayed yield response in an unconfined unit.

ison to the initial saturated thickness (Walton, 1970, p. 153).

Time-drawdown data are plotted on log-log paper and superimposed on a type curve such as Plate 8 in Lohman (1972). The parameters s and t are determined from a match point coincident with the dimensionless type curve parameters equal to one. Transmissivity and storage coefficient are then calculated from:

$$T = \frac{(1.0)\,Q}{4\pi s} \qquad (46)$$

and

$$S = \frac{4Tt}{r^2(1.0)} \qquad (47)$$

where T is the transmissivity (L^2/T), Q is the constant discharge rate (L^3/T), s is the drawdown (L), S is the early or late time value of the storage coefficient (dimensionless), t is the time since pumping began (T), and r is the radial distance to the observation well (L) (Lohman, 1972, pp. vi–vii, 36). The late time value of the storage coefficient approaches the value for the specific yield.

Neuman (1975) developed a solution for an anisotropic unconfined aquifer that considers the delayed response due to gravity drainage. The type curves are bounded by the Theis-type curve as are the previously described Boulton-type curves. Neuman (1975) assumed that the specific yield is much greater than the early time storage coefficient and that the drawdown is much less than the saturated unit thickness. The type curves presented by Neuman assume the pumping and observation wells are fully penetrating. Although not stated specifically, it can be assumed that the basic assumptions required by the Theis solution apply to Neuman's solution. Data are plotted on log-log paper as drawdown vs. time and superimposed on a type curve set prepared from tables presented by Neuman (1975). Match point parameters are first obtained for a late data match to the type B curves. Transmissivity is calculated from:

$$T = \frac{Qs_D}{4\pi s} \qquad (48)$$

where T is the transmissivity (L^2/T), Q is the pumping rate (L^3/T), s_D is the dimensionless drawdown, and s is the drawdown (L) (Neuman, 1975). Specific yield is calculated from:

$$S_y = \frac{Tt}{r^2 t_y} \qquad (49)$$

where S_y is the specific yield (dimensionless), t is the time since pumping started (T), r is the radial distance from pumping well (L), and t_y is the dimensionless time with respect to S_y (Neuman, 1975). Early time data are matched to the type A curves and match point values are again pulled from the data plot and the type curves. Transmissivity is calculated as before and this value should be equal approximately to the first value. Storage is calculated as before except that the equation is defined as:

$$S = \frac{Tt}{r^2 t_s} \qquad (50)$$

where S is the storage coefficient (dimensionless), and t_y is the dimensionless time with respect to S (Neuman, 1975). The other variables are as defined previously.

Transmissivity is defined as the radial hydraulic conductivity multiplied by the saturated thickness of the unit; therefore the radial and vertical hydraulic conductivities can be determined from:

$$K_D = \beta b^2/r^2 \qquad (51)$$

and

$$K_z = K_D\,K_r \qquad (52)$$

where K_D is the degree of anisotropy, β is the type curve match value (dimensionless), b is the initial saturated thickness (L), r is the radial distance from pumping well (L), K_z is the vertical hydraulic conductivity (L/T), and K_r is the radial hydraulic conductivity (L/T) (Neuman, 1975). Neuman (1975) states that the values obtained for transmissivity, storage coefficient, and specific yield are practically identical with those that would be obtained with Boulton's technique as described previously. Neuman's solution does permit the determination of vertical and horizontal hydraulic conductivity.

Fracture Aquifer Solutions—The nonsteady behavior of a confined aquifer during a pump test has been described by the Theis solution as discussed in the previous Nonleaky Artesian Aquifer Solution section. A significant variation in this response occurs if the well intercepts a horizontal or vertical fracture. Mathematical models developed by Gringarten (1974, 1974a) evaluate this response. These models and the significant findings of the model development were summarized by Gringarten and Witherspoon (1972) with respect to the analysis of confined aquifer test data.

Early time drawdown data from the pumped well were found to exhibit distinct characteristics when plotted on log-log paper (Fig. 46). An initial unit slope straight line plot occurs when the discharge from the well is derived from well storage that may be augmented by a large volume fracture intersected by the well (Ramey, 1975). An initial half-unit slope, straight line plot indicates the interception of a low volume fracture that creates linear flow to the well; the half-unit slope plot also will occur at observation wells if the same fracture is intercepted (Gringarten and Witherspoon, 1972). A data plot for a pumping well may reflect both discharge conditions by exhibiting an initial unit slope followed by a half-unit slope on the log-log plot. Data plots exhibiting a half-unit slope can be analyzed using the following models.

The vertical fracture model developed by Gringarten et al. (1974a) assumes that a vertical fracture totally penetrates a horizontal, homogeneous, and isotropic unit initially at a constant head. A single phase, slightly compressible fluid flows into the fracture at time zero, at a constant rate. The fracture is assumed to have an infinite hydraulic conductivity and the unit is assumed to be infinite in areal extent.

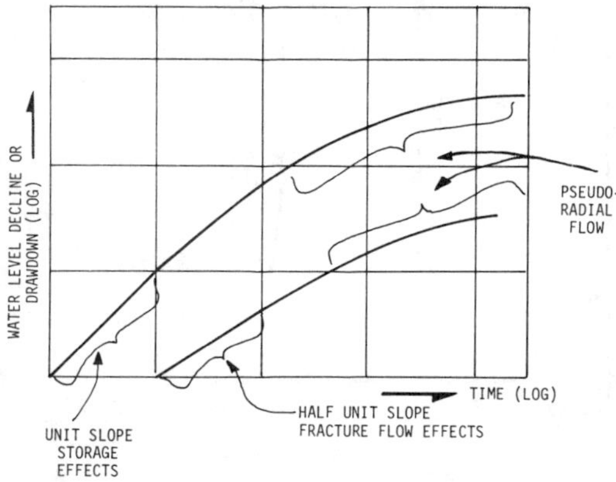

Fig. 46. Early time drawdown characteristics influenced by storage and fracture flow.

The horizontal fracture model (Gringarten and Ramey, 1974) assumes that the reservoir (unit) is horizontal, homogeneous, and anisotropic. The unit is penetrated fully by the well that intersects a single, horizontal fracture that is symmetrical and centered at the well. A slightly compressible fluid flows at a constant rate in the fracture as single phase flow. The inflow rate is uniform over the entire volume of the fracture. No flow occurs across the lower or upper boundaries (confining units) of the unit and the unit is infinite in areal extent.

A log-log plot of drawdown vs. time is superimposed on type curves prepared from tables presented by Gringarten (1974, 1974a). The type curve for a vertical fracture in an infinite reservoir is prepared using Table 1 and the column of values obtained for the uniform flux case as illustrated in Fig. 4 in Gringarten et al. (1974a). Type curves for the horizontal fracture case are prepared from tables by Gringarten and Ramey (1974). The appropriate type curve is fitted to the data and dimensionless pressure and time values determined for a match point corresponding to actual pressure and time data. A dimensionless unit thickness is also obtained in the horizontal fracture case.

Petroleum literature usually deals with intrinsic permeability rather than hydraulic conductivity. Intrinsic permeability is the property of the medium that permits the movement of a fluid but does not reflect the properties of the fluid. Hydraulic conductivity includes the properties of the medium to transport fluid and the properties of the fluid. The relationship between these two terms is:

$$K = \frac{k\rho g}{\mu} \qquad (53)$$

where K is the hydraulic conductivity (L/T), k is the intrinsic permeability (L^2), ρ is the fluid density (M/L^3), g is the acceleration due to gravity (l/T^2), and μ is the dynamic viscosity (M/LT) (Freeze and Cherry, 1979, p. 27).

The intrinsic permeability can be calculated in the vertical fracture case from:

$$P_{\omega D} = \frac{2\pi\kappa h}{q_f \mu}(P_i - P_{\omega f}) \qquad (54)$$

where $P_{\omega D}$ is the dimensionless wellbore pressure, κ is the intrinsic permeability (L^2), h is the thickness of the unit (L), q_f is the flow rate (L^3/T), μ is the fluid viscosity usually assumed to be one (M/LT) for water, P_i is the initial pressure (M/L^2), and $P_{\omega f}$ is the pressure after pumping or recovery has started (M/L^2) (Gringarten, et al., 1974a).

Transmissivity equals hydraulic conductivity multiplied by the thickness of the hydrostratigraphic unit. The length of the fracture can be estimated after determining the intrinsic permeability by utilizing the dimensionless time parameter:

$$t_D = \frac{\kappa t}{\phi \mu c x_F^2} \tag{55}$$

where t_D is the dimensionless time, κ is the intrinsic permeability (L^2), t is the flowing time (T), ϕ is the formation porosity (dimensionless), μ is the fluid viscosity usually assumed to be one for water (M/LT), c is the total compressibility of the unit (L^2/M), and x_F is the fracture half length (L) (Gringarten, et al., 1974a). Specific storage in petroleum literature is defined by $\phi \rho g c$ as used in Eqs. 58 and 55 (Narasimhan and Kanehiro, 1980).

The horizontal fracture equations are formed by replacing the κ in Eqs. 54 and 55 with κ_r, thereby denoting this as radial (horizontal) intrinsic permeability. The half fracture length (x_F) is replaced with r_F, the radius of the horizontal fracture. The final equations for a horizontal fracture become:

$$P_{\omega D} = \frac{2\pi \kappa_r h}{q_f \mu}(P_i - P_{\omega f}) \tag{56}$$

and

$$t_D = \frac{\kappa_r t}{\phi \mu c r_F^2} \tag{57}$$

where the other parameters are as defined previously (Gringarten and Ramey, 1974). An additional relationship, the dimensionless reservoir (unit) thickness, is defined as:

$$h_D = \frac{h}{r_F} = \sqrt{\frac{\kappa_r}{\kappa_z}} \tag{58}$$

where h_D is the dimensionless reservoir thickness, h is the unit thickness (L), r_f is the fracture radius (L), κ_r is the radial intrinsic permeability (L^2), and κ_z is the vertical intrinsic permeability (L^2) (Gringarten and Ramey, 1974). The vertical intrinsic permeability can be calculated using Eq. 58 after calculating the radius of the fracture.

The reader should note that by the time this volume went to press in February 1986, several important new papers had appeared on this subject. It is not possible for us to incorporate them herein.

Vertical Hydraulic Conductivity—The vertical hydraulic conductivity of a unit can be "estimated" from a single well. Fluid is injected or removed from one zone of the unit while pressures (heads) are monitored in a separate isolated zone as illustrated in Fig. 47. The method of analysis developed by Prats (1970) assumes that the unit being tested is infinite in areal extent, has a uniform thickness and porosity, is homogeneous, and can be isotropic or anisotropic. The anisotropy is considered along a vertical axis. The fluid filling the unit has a constant compressibility and viscosity at an initially uniform pressure or head.

Fluid is injected or removed at a constant rate through one of the casing ports or perforations while the pressure or head change is monitored with time in the other port. Fluid is *not* added or removed from the second port. The pressure (head) change is plotted vs. time on semi-log paper as illustrated by Prats' Fig. 3 (1970). A conventional

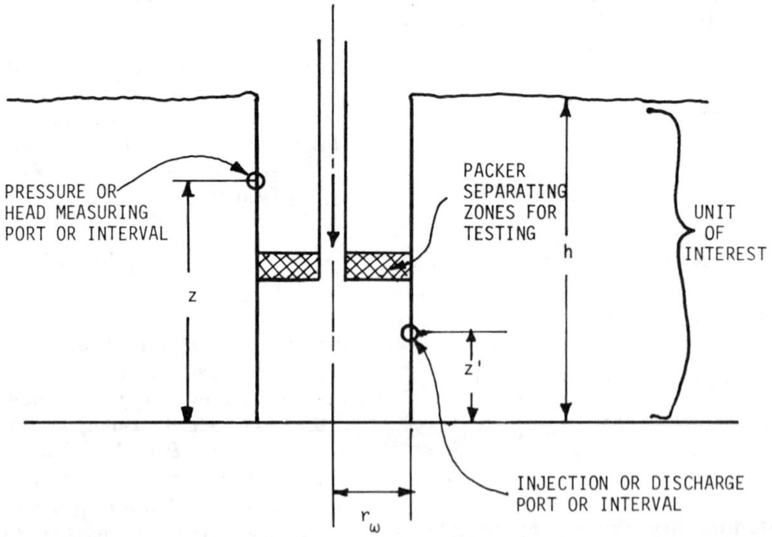

Fig. 47. Single well test for vertical hydraulic conductivity (modified after Prats, 1970).

semi-log analysis can be carried out to determine the transmissivity of the unit. The vertical intrinsic permeability is calculated from:

$$\kappa_v = 3788\ \phi\mu c h^2 \exp\left[\frac{2.303\ \Delta P(1)}{\Delta P/1\ \text{cycle}} - F(z/h,\ z'h)\right] \quad (59)$$

where κ_v is the vertical intrinsic permeability (millidarcies), ϕ is the porosity (dimensionless), μ is the viscosity (centipoise), c is the total compressibility (psi^{-1}), h is the unit thickness (in feet) (L), $\Delta P(1)$ is the extrapolated straight line segment of pressure change at one hour (psi), $\Delta P/1$ cycle is the pressure change per log cycle (psi), and $F(z/h,\ z'/h)$ is the geometrical function determined from Fig. 2 as a function of vertical boundaries and locations of ports from Prats (1970). Specific storage in petroleum terms is $\phi\rho gc$ as noted previously.

Obviously, cased wells being tested through perforations at different elevations must be properly cemented to prevent vertical fluid migration between the borehole and casing. Gravel packed wells are not suited for this test. The injection and measuring interval should probably be less than 10% of the distance between the ports or intervals. The distance between the ports should be on the order of 25 well radii times the anticipated ratio of vertical to horizontal hydraulic conductivity (Prats, 1970).

Any single well vertical permeability test suffers from the risk that the test may measure the properties of the surrounding rock that has been disturbed by the drilling and the borehole fluid.

Pump Test Recovery—Constant rate pump tests can provide two sets of data for determining hydrogeologic properties. The initial drawdown data can be evaluated as described previously and the recovery data after the pump is shut off can be analyzed also. The recovery of the water level in the well can be visualized by superimposing an injection well with an injection rate equal to the original discharge rate. The recovering water level appears as an inverted image of the declining water level from the pump test as illustrated in Fig. 48.

Recovery data are plotted as recovery (Fig. 48) vs. time since pumping stopped. Alternatively residual drawdown can be plotted vs. the ratio of time since pumping began to time since pumping stopped. The application of recovery data to the constant discharge tests previously described is bounded by the constraints stated with each method.

Boundary Effects—The solutions described previously for pump tests assume that the unit being pumped is infinite in areal extent. No hydrostratigraphic unit is truly infinite but the response to pump tests may approximate a unit that is infinite

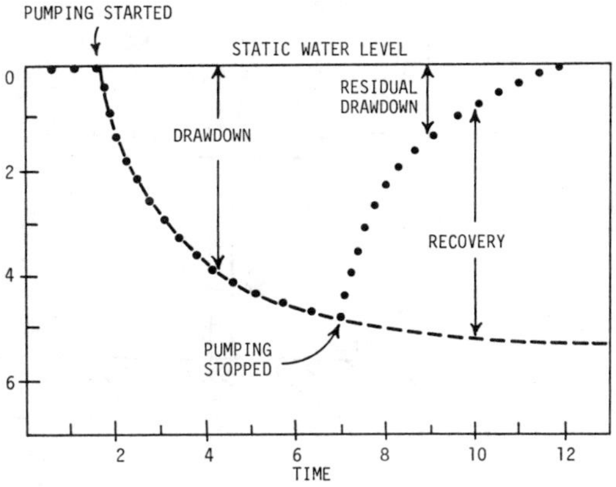

Fig. 48. Typical drawdown and recovery curves for a constant rate pump test (modified after Johnson Division, UOP, 1972, p. 136).

if the wells are located sufficiently far from the real boundaries. Boundaries can be classified as recharge boundaries or barrier boundaries. The effect created by boundaries can be simulated hydraulically by an image well or a series of image wells, depending upon the configuration of the boundaries. An image well would be an injection well for a recharge boundary or a discharge well for a barrier boundary. Fig. 40d illustrates the effects of boundaries on pump test data. The simplest configuration consists of a linear boundary as illustrated in Figs. 49 and 50.

Distances to the image well can be calculated from pump test data. The actual boundary is located halfway between the image well and the discharge well. Data from three observation wells are required to determine the location of a boundary.

Stallman (1963) developed a procedure for calculating the transmissivity, storage coefficient, and distance to an image well by modifying the Theis nonleaky artesian solution. A series of type curves were developed by Stallman (1963) and appear in his Plate 3 and in Lohman (1972, Plate 9). Drawdown vs. time in an observation well is plotted on log-log paper and superimposed on the type curves. Match point values are obtained and inserted into:

$$T = \frac{Q}{4\pi s_0}\ \Sigma W(u) \quad (60)$$

and

$$S = \frac{4Tt}{r_p^2}\ u \quad (61)$$

where T is the transmissivity (L^2/T), Q is the constant discharge rate (L^3/T), s_0 is the algebraic sum

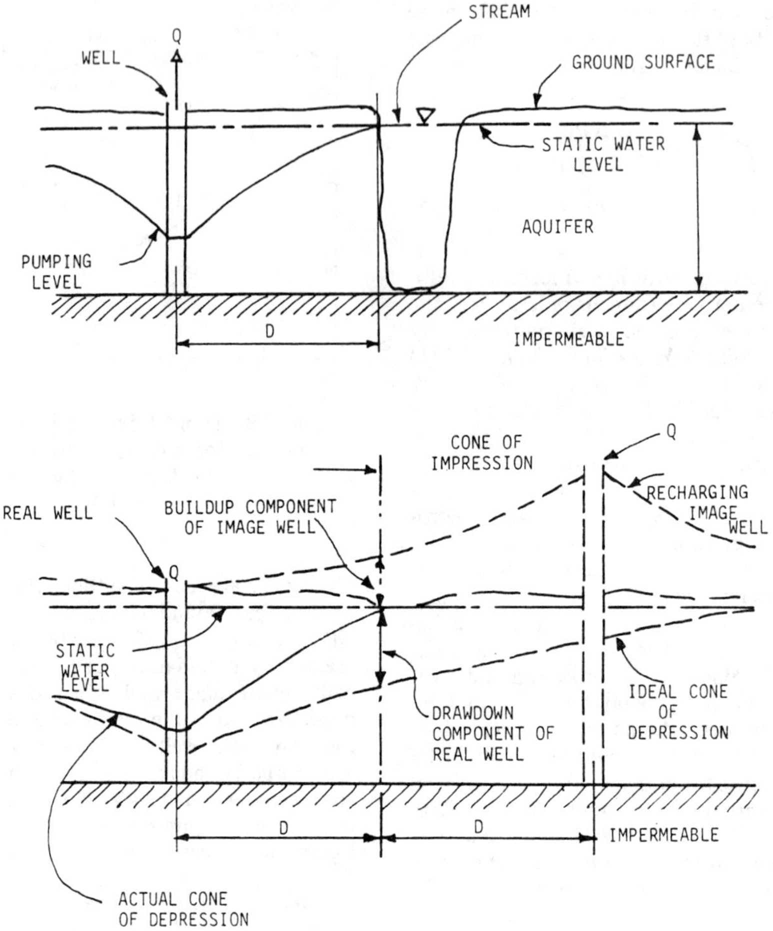

Fig. 49. Relationship of a recharge boundary and an image well (US Bureau of Reclamation, 1977, Fig. 5–18).

of drawdown due to pumping and image wells (L), $\Sigma W(u)$ is the simulation of $W(u)$ values for pumping and image well (dimensionless), S is the storage coefficient (dimensionless), t is the time since pumping began (T), r_p is the radial distance from observation well to pumping well (L), and u_p is the variable of integration (dimensionless). A κ value is also determined from the type curve match and inserted into:

$$\kappa = \frac{r_i}{r_p} \tag{62}$$

where κ is the constant of proportionality (dimensionless), r_i is the radial distance from observation well to image well (L), and r_p is the radial distance from observation well to pumped well (L) (Lohman, 1972, pp. 60–61).

The distance from the observation wells to the image well can also be determined by utilizing either the log-log plot or the semi-log plot of drawdown versus time and the law of times. The law of times states that:

$$\frac{r_i^2}{t_i} = \frac{r_p^2}{t_p} \tag{63}$$

where r_i is the radial distance from observation well to image well (L), t_i is the time after pumping has started when the divergence of the actual time drawdown plot from the type curve or ideal drawdown plot under the influence of the boundary (image well) is equal to the particular value of drawdown at t_p (T), r_p is the distance from pumping well to observation well (L), and t_p is the time after pumping has started for some particular value of

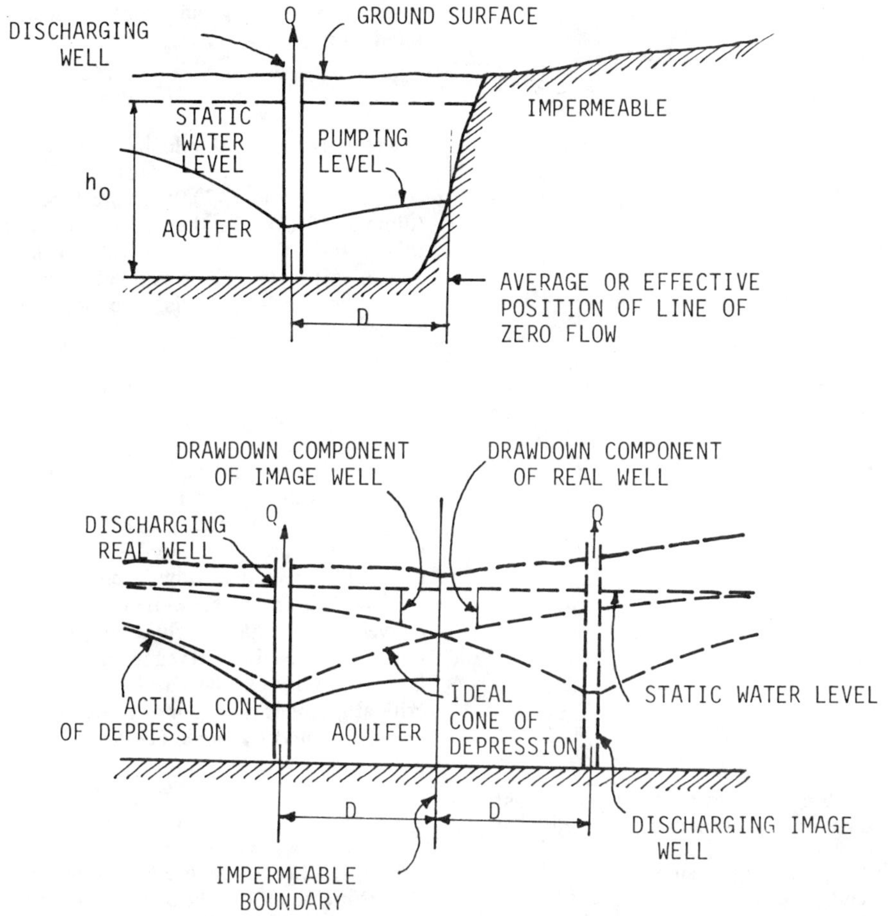

Fig. 50. Relationship of an impermeable boundary and an image well (US Bureau of Reclamation, 1977, Fig. 5–17).

drawdown unaffected by boundaries (T) (Walton, 1962, 1967, p. 16). Time drawdown data are plotted as shown in Fig. 51a or 51b. The effect of the boundary as an image well is determined for some time t_i, as an increment of drawdown, s_i. An equal value of drawdown, s_p, is located on a portion of the drawdown plot unaffected by the boundary. The time, t_p, is noted. The radial distance to the image well can be calculated by substituting into Eq. 63.

Boundary conditions can be difficult to interpret solely from drawdown data as shown by Fig. 40a-c. Geologic and geophysical data can be a valuable asset in evaluating pump test data; in many cases these data are a necessity for proper interpretation.

Geophysical Techniques—In some cases the vertical hydraulic conductivity of a unit can be estimated through the use of ground water temperatures. Bredehoeft (1965) developed such a procedure for using the thermal gradient in a semiconfining unit that is isotropic, homogeneous, fully saturated, and experiencing steady state vertical flow. Fluid temperature is measured at three or more points along a vertical section of a semiconfining unit. The hydraulic gradient across the unit should also be measured as noted in Fig. 52. The ratio of the temperature data is plotted vs. the depth factor as:

$$(T_z - T_0)/(T_L - T_0) \qquad (64)$$

and

$$2/L \qquad (65)$$

where T_z is the temperature at depth z, T_0 is the uppermost temperature, T_L is the lowermost temperature, z is the distance positive downward from T_0 to T_z (L), and L is the length of vertical section

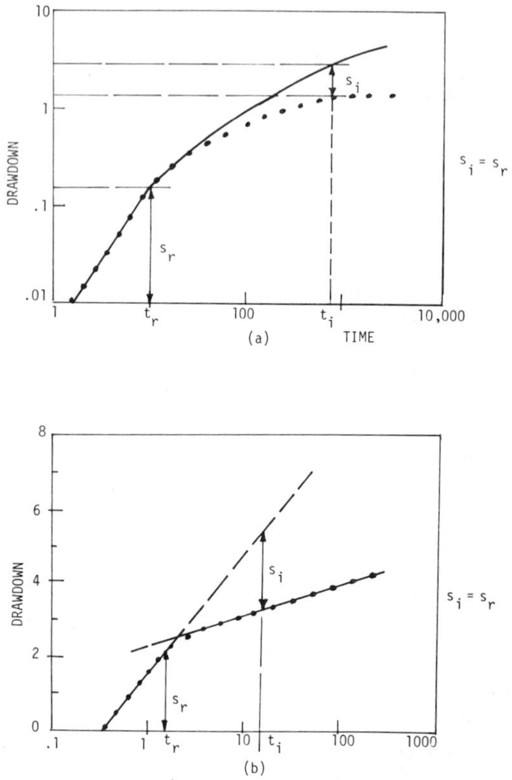

Fig. 51. Time drawdown plots of pump test data for determining distances to boundaries.

over which temperature measurements were made between T_0 and T_L (L).

Plotted data are superposed on a set of type curves prepared from tables presented by Bredehoeft (1965). The ground water velocity is calculated from:

$$v_z = \kappa\beta / c_0\rho_0 L \tag{66}$$

where v_z is the fluid velocity in vertical direction (L/T), κ is the thermal conductivity of solid-fluid complex (calorie/L-T), β is the type curve value (dimensionless), c_0 is the specific heat of fluid (calorie/gram—°C), ρ_0 is the density of fluid (M/L^3), and L is the length of vertical section (L). The hydraulic conductivity is then obtained by using Darcy's law (Eq. 1) as:

$$|v_z| = \left|\kappa \frac{dh}{dz}\right| \tag{67}$$

where v_z is the fluid velocity in vertical direction (L/T), κ is the hydraulic conductivity (L/T), and dh/dz is the hydraulic gradient along the vertical thermal section (dimensionless). The technique is applicable to vertical flow in either direction.

The fluid in a single well must be stable, without convection, so that the fluid temperature is representative of the thermal regime in the semiconfining layer. Stability can be checked by calculating the critical gradient that cannot be exceeded. This gradient, as noted by Boyle (1979), is:

$$G_c = \frac{ga\phi}{c} + \frac{cv\alpha}{gar^4} \tag{68}$$

where G_c is the critical gradient at which convective flow is incipient, g is the acceleration due to gravity (cm/s^2), a is the volume coefficient of thermal expansion (1/°C), ϕ is the absolute temperature

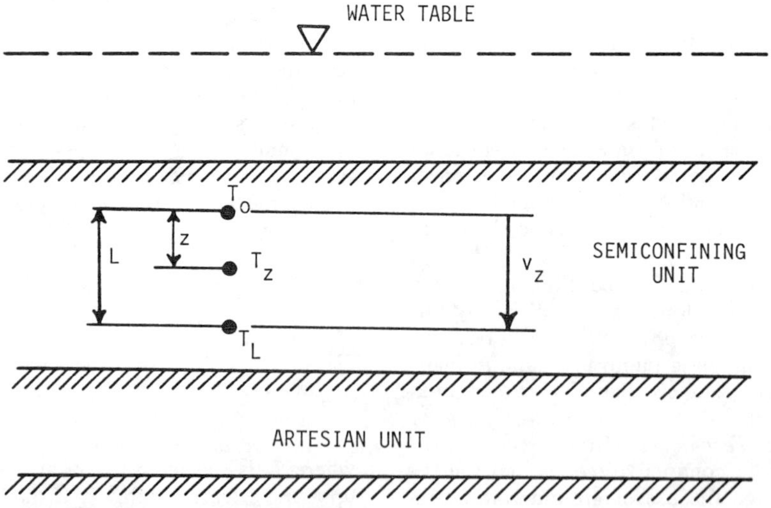

Fig. 52. Measured temperature for determining vertical hydraulic conductivity (modified after Bredehoeft and Papadopulos, 1965).

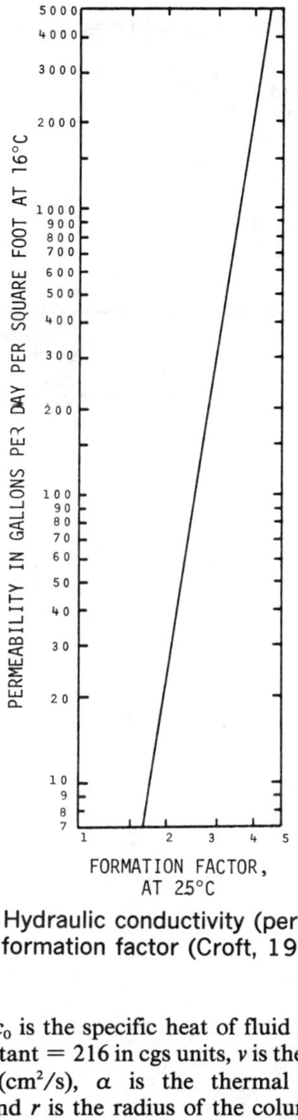

Fig. 53. Hydraulic conductivity (permeability) vs. formation factor (Croft, 1971).

(Kelvin), c_0 is the specific heat of fluid (ergs/g°C), C is a constant = 216 in cgs units, v is the kinematic viscosity (cm^2/s), α is the thermal diffusivity (cm^2/s), and r is the radius of the column (cm).

A method is available for "estimating" the hydraulic conductivity (permeability) of fresh water materials with borehole electric logging devices. The method has been field tested by Alger (1976), Croft (1971), and Jones and Buford (1951), utilizing core samples and aquifer tests to show that the technique can be accurate to within 10%. The key to the technique is the relationship between the formation factor and hydraulic conductivity (permeability).

The formation factor (F) is the empirical relationship between the water and rock resistivities:

$$F = \frac{R_0}{R_\omega} \quad (69)$$

where F is the formation factor, R_0 is the resistivity of the water-saturated rock, and R_ω is the resistivity of the water with which the rock is saturated (Alger, 1976). Values of R_0 for the subsurface formations are determined from the long normal borehole resistivity log that can be run by borehole geophysics service companies. R_ω for the water must be taken from either in situ conductivity or resistivity readings or by testing water samples taken from the borehole.

The relationship between F and hydraulic conductivity (permeability) is illustrated in Fig. 53 that was determined from data relating grain size to permeability and formation factor by Croft (1971). The formation factor for any hydrostratigraphic unit can be entered into the graph to yield a hydraulic conductivity (permeability) in gallons per day per square foot at 16°C. Radiation and drillers' logs can be used to delineate aquifer thickness. Transmissivity can be obtained by multiplying the unit thickness by the appropriate hydraulic conductivity (permeability). Table 3 is an example of this procedure. This procedure should be viewed as yielding only an estimate of transmissivity; it should be supported by other methods.

Computer Techniques—Computer modeling techniques have been used to determine the transmissivity and storage coefficient of hydrostratigraphic units. Land (1977) utilized a trial-and-error approach for matching computer generated drawdown data to actual drawdown data collected during a pump test of a multilayered system. The hydrogeologic system was complex and not readily simulated by conventional methods of analysis. The simulation utilizes a wedge-shaped section of the total hydrogeologic system with the apex of the wedge passing through the pumping well. The hydrogeologic parameters are adjusted by trial and error until the simulated drawdown data match the actual drawdown data.

Hydrogeologic properties can also be estimated by using a technique referred to as the inverse problem or the direct approach. Unfortunately, a solution of the inverse problem is not unique (Freeze and Cherry, 1979, p. 359). There are many unknowns in the solution. A recent approach to the inverse problem has been proposed by Neuman (1980). This is a statistically based approach to estimating spatially varying transmissivities on the basis of steady water level and flux data. The method is utilized by a finite element scheme and log transmissivities. The log transmissivity estimates are only mildly influenced by small errors in the model and in the flow rate and sink/source data, but low-amplitude noise in the water level data may cause the log transmissivity estimates to become unstable.

Table 3. Determining Transmissivity Values from Borehole Geophysical Logs (Modified after Croft, 1971)

Hydrostratigraphic unit (depth interval, feet)	R^*_0 in ohm-meter²/meter (from long normal log)	R^*_ω in ohm-meter from in situ or sample determination	F dimensionless
1190–1210	6.2	3.5	6.2/3.5
1250–1280	6.4	3.5	6.4/3.5
1310–1350	5.8	3.5	5.8/3.5
1430–1455	6.4	3.5	6.4/3.5
1480–1505	5.8	3.5	5.8/3.5

Hydraulic conductivity (permeability), gal per day per sq ft	Unit thickness, feet (from geophysical logs or well logs)	Transmissivity of individual units, gal per day per ft
12	20	240
12	30	360
9	40	360
12	25	300
9	25	225
Overall transmissivity:		1485†

* These must be converted to the same temperature; use Schlumberger Technology charts by Schlumberger (1968).
† When this was determined from a conventional pump test, the value obtained was 1400 gal per day per ft.

5

Application of Mathematical Models to Mine Water Inflow

The literature on four different topics is reviewed herein for purposes of mathematical modeling of mine water inflow. The first of the topics reviewed is modeling of fissured or fractured materials; the second is general Darcy based flow models; the third is field methods for determining fracture flow properties; and the fourth is methods used presently for determining mine water inflows. The four topics are considered in the foregoing order.

FRACTURE FLOW MODEL

Two basic types of fracture flow models have been developed. The first type uses various techniques for space averaging to obtain the equations. Systems are considered in which the porous media properties vary smoothly enough and the fracturing is sufficiently profuse and well distributed that the locally averaged properties are meaningful when viewed on the macroscopic scale. The final result is that a coefficient is used which is similar to saturated hydraulic conductivity and in many cases, simply is hydraulic conductivity. In other cases, the coefficient is apparent hydraulic conductivity because the response from flow in the fractures is more rapid than flow in the pore spaces. Faust and Mercer (1980) refers to this method as the continuum approach. Some of the papers on this type of problem are by Duguid (1973), Verner et al. (1974), Owili-Eger (1975), Duguid and Abel (1974), duPrey and Weill (1974), and O'Neill (1978). Duguid and Lee (1977) also considered heat flow in the fluid and the porous medium. These papers all result in various types of computer programs. Some papers on this topic that do not result in computer programs are by Snow (1968, 1969) and Streltsova (1976). Some of the foregoing articles consider consolidation of the porous media and the resulting effect on hydraulic conductivity. Barenblatt et al. (1960) developed an analytical solution for fracture flow and discussed the time that is involved in the drainage process and the effect that fracture flow has on the drainage time.

The second type of fracture flow model considers the width of the fractures as a function of the pressure in the field. Faust and Mercer (1980) refers to this method as the discontinuum approach. This approach often requires two finite element programs that consider the stress in the material and the flow in the fractures separately. The programs are then linked to arrive at the final solution. Some of the articles that consider the interrelationship between pressure and the size of the fracture but do not use computer solutions are Gringarten et al. (1975), Narasimhan and Palen (1979), and Witherspoon et al. (1979). Papers that use a linked computer program between stress and flow are by Ayatollahi (1978), Gale et al. (1974), and Noorishad et al. (1971). Ayatollahi's program is primarily for the petroleum industry because it does not consider the effect of gravity. Ayatollahi assumes that pressure effects are very large in comparison to gravity effects. This assumption would apply only to very deep aquifers.

Articles that consider fracture flow on a more applied basis are by Wittke et al. (1972), who considers that the flow is entirely through fissures and negligible through the pores of the rock. Gringarten and Witherspoon (1972) developed type curves that may be used for analysis of fracture flow systems; the curves are based on the assumption that it is possible to distinguish between aquifers with horizontal and vertical fractures and that it is possible to analyze the system as an equivalent anisotropic homogeneous porous medium with a single fracture with much higher permeability.

DARCY FLOW MODELS

The Darcy flow models of greatest interest to mine water inflow prediction are those which allow for unsaturated flow. The consideration of unsaturated flow is important in the analysis of mine inflow because perched water tables are allowed to develop if layered materials of variable hydraulic conductivity occur. Many of these programs have resulted from work at the University of California at Berkeley by a number of different authors. Some of these papers are by Narasimhan (1977, 1978a, 1978b) (who also considers the consolidation of the material) and Narasimhan et al. (1978) and Neuman (1974).

A three-dimensional model which has been used but not verified was developed by Frind and Verge (1978). Sagar (1978) treats the properties of the porous media as random variables and uses a Galerkin finite element formulation but does not present a computer program. Chowdhury (1974) considers a generalized steady state field problem.

Other computer models which have been used for determining inflow to mines are presented by Ricca and Hemmerich (1978), Schubert (1978), and Szilaggi et al. (1978). Ricca used the Stanford watershed model and determined mine inflow from precipitation and streamflow records. These three papers are not finite element nor finite difference based models; they basically are water balance models.

Innumerable other ground water models based on the Darcy equation appear in the literature. Several of these are three dimensional in the sense that they consider flow in various layers of a hydrostratigraphic section and simulate flow between the aquifers in the section. This characteristic would be necessary in considering mine water inflow if unsaturated flow is not considered.

PROPERTIES OF FRACTURED MATERIALS

Articles from the petroleum industry are directed primarily toward determination of material properties from borehole information: Examples of such articles are by Patton (1979), Prats (1970), Ramey (1970), Johnson et al. (1966), Russell (1966), Brigham (1970), McKinley et al. (1968), Millheim and Cichowicz (1968), Kamal and Brigham (1976), Crawford et al. (1976), Cinco et al. (1975), Aguilera (1976), Jargon (1976), Agarwal et al. (1970), Earlougher and Kersch (1974), Wattenbarger and Ramey (1969), Gringarten et al. (1974a), Gringarten and Ramey (1974), Chu et al. (1980), and Earlougher (1980). Several of these articles are discussed in greater detail elsewhere in this volume.

EXISTING MINE INFLOW TECHNIQUES

Computational techniques that have been developed exclusively for the purpose of simulating mine water inflow rates are limited in number. The most comprehensive analysis is by Brown (1977) who developed such a treatment for the water management plan for a Piceance Basin oil shale project in Colorado. Oil shale mines are to be developed on a room-and-pillar basis with a mined out area somewhat circular in shape, 60 to 100 ft (18–30.5 m) high, and a mile or more in diameter.

Brown (1977) separated the flow into the mined region into three parts: (1) the flow from the top of the mine; (2) flow from the sides, assumed to be radial; and (3) flow from the bottom. These zones are shown in Fig. 54. For the flow in part 1 a finite element solution was used in conjunction with a generalized flow net and several dimensionless plots. The steady state flow from this region was determined with a fixed water table. Other flow rates that were considered were those that would occur if a fixed head existed at the far boundary or if no flow occurred at the far boundary. These solutions all were presented in terms of a dimensionless time parameter. For flow from the periphery of the mine (part 2), a modified form of the unsteady well flow equation (Theis equation) was used. Modification was necessary because the radius of the well (mined area) was expanding and no depth of water exists in the well; that is, the entire wall of the well would be a seepage surface. In zone 3 flow was assumed horizontal below the level of the mine and vertical as it reached the region beneath the mine.

Many different mining configurations were investigated by Brown. These varied from considering just the mined out portion of the Mahogany zone (the highest quality oil shale) to considering also the caved portion from the mine to the ground surface, a distance of nearly 1000 ft (305 m). A mine was also considered in what is called the R-4 zone that is about 1600 ft (488 m) below the water table or 1800 ft (549 m) beneath the ground

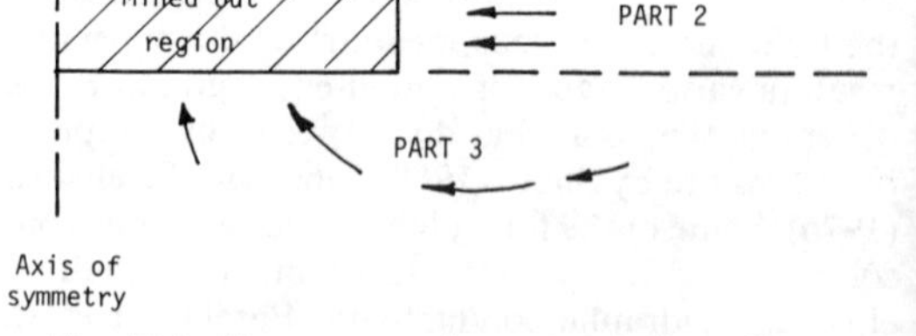

Fig. 54. Flow zones for mine inflow (from Brown, et al., 1977).

surface. A total of 112 different boundary configurations was used. The calculated values of inflow ranged upward to as high as 56,000 gpm (212 m^3) for the maximum rate of expansion of the mine.

Brown (1979) presented another analysis in which he made use of the unsteady well flow (Theis) equation and the Hantush and Jacob leaky-aquifer equation to determine the inflow rates as a vertical shaft would be sunk in the Piceance Basin to a depth of 2000 ft (610 m). He stated in this paper "for oil shale, geo-hydrology analysis the use of simple models and simple analytical relationships produces results which are the degree of accuracy which is appropriate to the accuracy of the parameters and in general to the needs of mine planners." None of Brown's work considered inflow to a horizontal drift.

Worley (1962) determined inflow to a vertical shaft by using the Theis equation with a modification to account for the effect of varying hole diameter. The equation that he used for the effect of the varying diameter was simply the steady flow well equation. To determine whether the methods of Brown and Worley agree we have solved a sample problem using each of the methods. Our calculations are given in Appendix B. The results of the two analyses do agree reasonably well. The inflow, using the Brown method, was 168 gpm (0.6 m^3/min) whereas the inflow using Worley's method was 199 gpm (0.8 m^3/min). This value depends to a great extent on the size of well used to obtain the field data for the sample problem. These two articles are the only articles we have encountered that consider the unsteady flow into a vertical shaft where the entire face of the shaft is a seepage surface.

Brown also computed the rate of inflow into a 24-ft (7.3-m) diam vertical shaft as it progressed at a rate of 5 ft (1.5 m) per day. He assumed that the shaft was sealed except for the bottom 50 ft (15 m). Although he states that a leaky aquifer analysis was used, it is not clear whether he accounted for all the various layers as the shaft progressed to a depth of 1700 ft (518 m). It is doubtful that he accounted for more than two layers at a time. He apparently assumed also that the water table remained constant at 330 ft (100 m) as he calculated inflows at various depths. Fig. 55 shows a plot of Brown's calculated inflow as a function of depth.

We have used the finite element program UNSAT2 to simulate the identical problem. The finite element mesh we used is shown in Fig. 56. The mesh portrays an axial symmetric problem with the vertical axis of symmetry at the left side. To start the simulation, the radial coordinates of all but the top two nodes were set at zero. The radial coordinates of the top two nodes were set at 12 ft (3.7 m) to represent a shaft 24 ft (7.3 m) in diameter. These points represent the boundary of the shaft when it was at a depth of 50 ft (15 m). They were also set as a seepage surface. They were allowed to drain for a period of 10 days, the time required for the shaft to advance a distance of 50 ft (15 m). The simulation then stopped; the next node was moved to a radius of 12 ft (3.7 m); and the top node was specified as a no flow node. Simulation again proceeded for a 10-day period at which time the shaft and seepage surface were again advanced 50 ft (15 m). This process was repeated until the shaft was at a depth of 1200 ft (366 m) below the original water table. This depth constituted an actual depth of 1530 ft (466 m) because the 330 ft (100 m) above the water table were not included in the finite element mesh.

The average inflow rates produced by the simulation are also shown in Fig. 55. The variation of flow from our simulation agrees very well with the variation in Brown's calculated values, but our actual values are somewhat larger than Brown's. There are several possible reasons for this discrepancy: First, it is not certain that Brown's analysis is correct since he had to make several modifying assumptions in the application of the well equations. Second, when our simulation was run the two nodes for the inflow portion of the shaft were 50 ft (15 m) apart so as to represent the 50 ft (15 m) of open shaft. However, we realized later that this procedure left approximately 100 ft (30.5 m) of shaft open because the surface represented by each node extends half way to the next node. To determine how much effect this would have on inflow, we repeated the solution for a depth of 1150 ft (350.5 m) with the nodes at the proper spacing for representing 50 ft (15 m) of open shaft. The inflow was reduced to about 60% of our original value. This point is plotted in Fig. 55. A reduction to 60% of the plotted line would agree very closely with Brown's values. Third, it is not completely clear that Brown always had 50 ft (15 m) of shaft open for a 10-day period because after stating that the shaft advance rate was 5 ft (1.5 m) per day and that the open shaft length was 50 ft (15 m), he stated that the average time required for the open shaft was five days. This discussion is presented to illustrate the modeling complexities that can influence mine water inflow predictions significantly.

The water table elevation during the progress of the shaft is noteworthy. During the first several time periods the water table dropped about 15 ft (4.6 m) near the shaft. Later on as the seeping portion of the shaft dropped in elevation, an actual rise of the water table occurred near the shaft and the drawdown extended over a wider region. Fig. 57 shows the water table location for shaft depths of 50, 600, and 1200 ft (15, 183 and 366 m). At no time did unsaturated regions and perched water tables form below the general water table.

In reality, it appears that the comparison be-

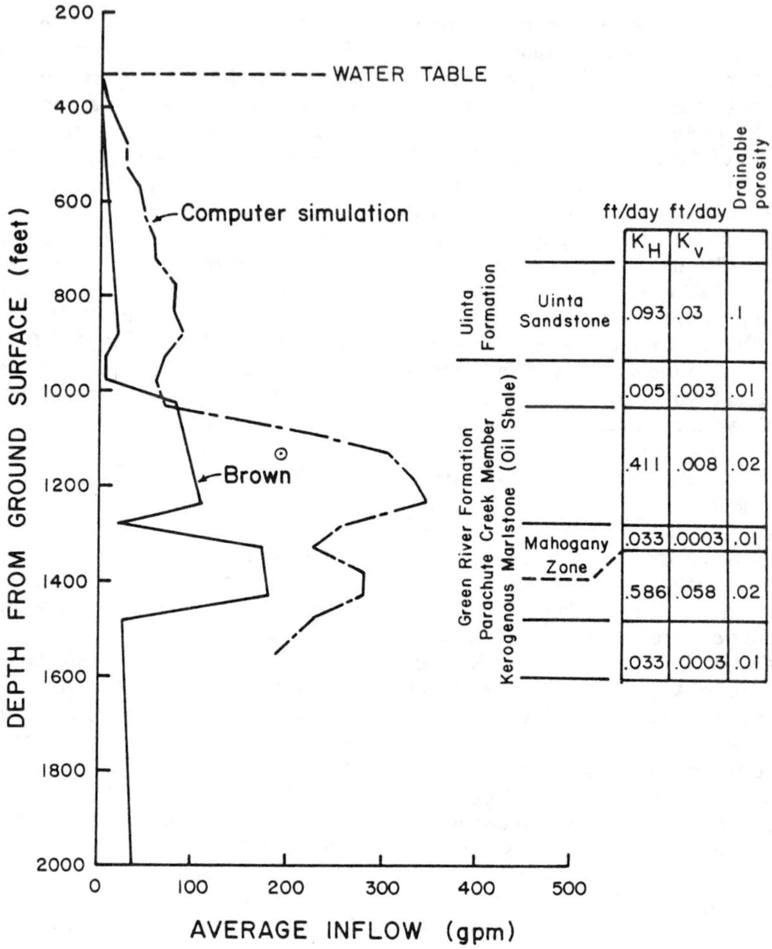

Fig. 55. Inflow to vertical shaft, Piceance Basin, Colorado (modified after Brown, 1979, Fig. 5).

tween the two methods constitutes somewhat of a verification for both methods since the variations in inflow rates are quite similar. Our computer simulation would be more expensive than Brown's simplified analysis. Approximately 18 hr of computer time and about 20 man-hr of our time were required for our complex analysis. However, a great deal more information can be obtained from our simulation because pressures at any location and time are obtained that could be of great importance in determining how to seal the shaft. The effect of other boundary conditions in the shaft also could be determined easily.

An analysis for flow into a horizontal drift is given by Freeze and Cherry (1979, pp. 490–491). This equation is attributed to Goodman (1965) who obtained it from Polubarinova-Kochina (1962, p. 374). The equation that was developed by use of complex variables and conformal mapping gives inflow rate to a drift in terms of hydraulic conductivity, radius of the drift, and vertical distance from the drift to a constant water table. Polubarinova-Kochina (1962, pp. 565–570) also developed an equation for unsteady flow where the water table above the drift is dropping. The initial discharge for this case is the same as the constant rate discharge. We consider these equations in detail subsequently when methods are developed for determining inflows by hand calculation procedures. Pauker (1975) presents several equations concerning the flow into horizontal drifts. Several of these equations give the rate of drop of the piezometric surface above the drift while the others give the discharge into the drift. These are the only methods we have encountered in the literature by which inflow to drifts (tunnel) or shafts of any configuration may be determined.

The scarcity in the literature of relatively simple methods for determining mine water inflow as opposed to computer models that apparently could

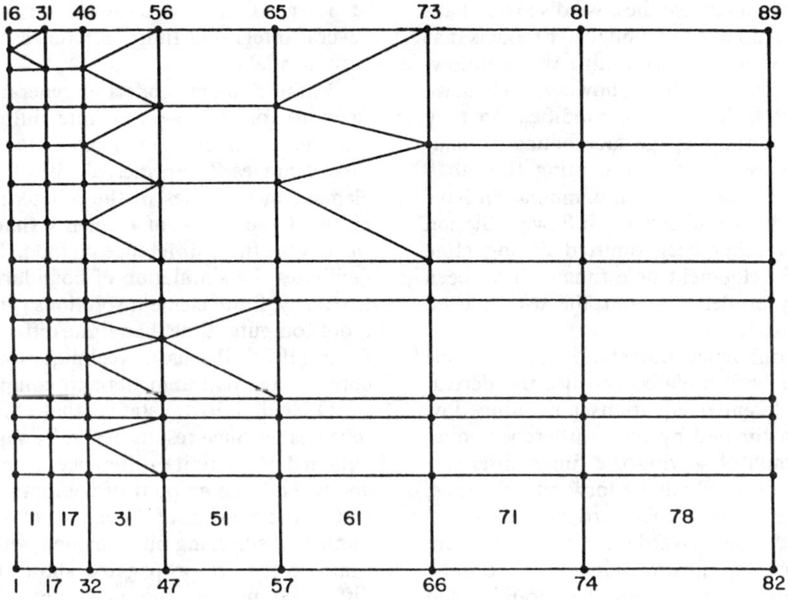

Fig. 56. Finite element mesh for a vertical shaft.

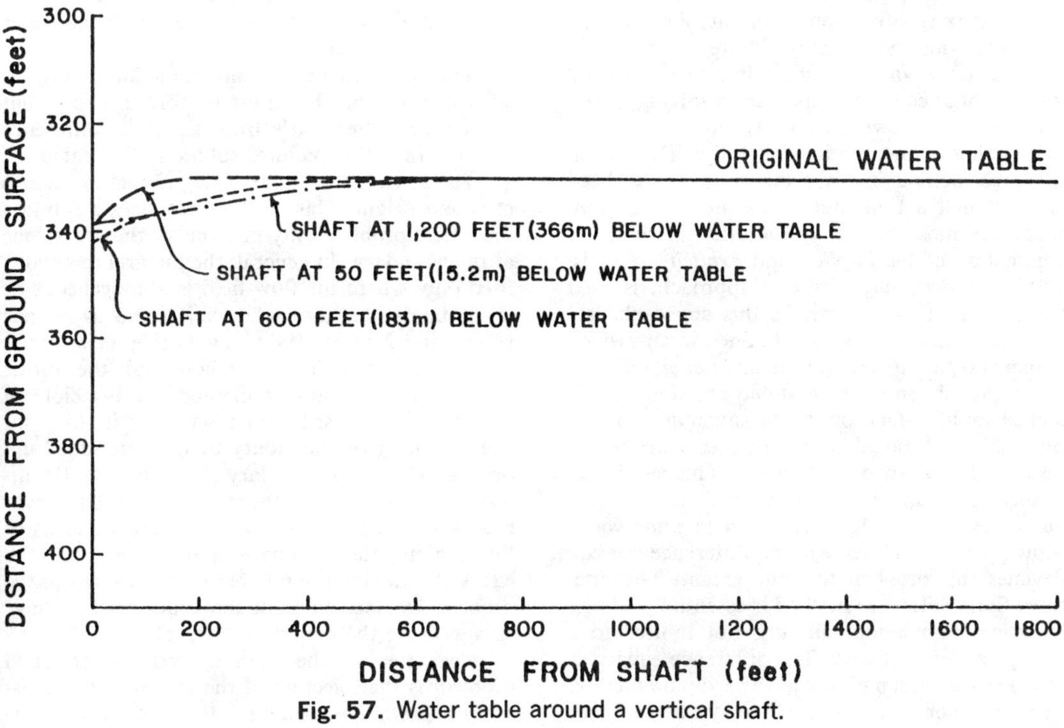

Fig. 57. Water table around a vertical shaft.

be used for determination of inflow is noteworthy. After several computer searches, we discovered approximately 300 different ground water models that perhaps could be adapted to a mine water inflow. We abandoned most of these, however, when we determined that without major modification they could not handle the seepage face boundary condition that is necessary for simulating flow to a mine shaft. These discoveries in combination have led us to conclude that mine water inflow prediction historically has either been omitted during mine planning and development or estimates have been conducted very crudely without the use of even hand calculations.

The finite difference numerical model is the model most easily formulated because the derivatives in the differential equations governing flow are easily approximated by finite difference equations. The method of solving the finite difference equations depends on whether a forward difference, a central difference, or a backward difference formulation is used. The forward difference procedure involves determining future values of hydraulic head in terms of existing values of hydraulic head at the nodes. This procedure gives a fully explicit equation for future heads that is solved very easily. However, this formulation experiences stability problems in that if the time steps are too large a stable solution will not be attained. The backward difference procedure results in an *implicit* equation at each node that has several unknowns. The size of the matrix is a function of the number of node points and vice-versa. Many different procedures are used for solving a matrix, but in general all involve longer computer times than solving an *explicit* equation. However, an advantage of the explicit method is unconditional stability. The central difference method involves determining the head values at half a time step. This method is more accurate in most cases than the other methods. A combination of the *implicit* and *explicit* methods, called the *alternating direction* approach, is used widely at the present time. In this approach part of the node point values are obtained in an *explicit* manner and part are obtained in an *implicit* manner. This approach produces a stable solution that is reached rapidly. One of the disadvantages of the finite difference model is that it is necessary to use a rectangular system of node points. This restriction produces difficulty in fitting a mesh to irregular boundaries such as those common to mine water inflow problems. The integrated difference method alleviates this problem to some extent. The integrated finite difference method is essentially a finite difference formulation but it is not limited to a rectangular grid network. The difference equations are written in the same form as the ordinary method except that more than four nodes may exist adjacent to a particular node. Consequently, various types of polygons may be used to construct the mesh. Several integrated finite difference codes are presently available.

Finite element models in general are more difficult to formulate than finite difference models, but the advantages in many cases outweigh the disadvantages. The principal advantage of the finite element method lies in the inherent greater flexibility in the shape of elements that can be used than in the finite difference method. This advantage facilitates the simulation of boundaries. Disadvantageously, finite element solutions generally require more computer time than finite difference solutions. Luhr (1974) discusses techniques for improving core storage requirements and computer time.

O'Neill (1978) states that "finite element schemes produce results generally superior to those obtained from finite difference schemes involving about the same amount of computational labor. A systematic analysis of the nature of oscillations and numerical smearing phenomenon which arise when sharp fronts are propagated shows that the finite difference methods are consistently inferior to the finite element schemes investigated. Also the finite element method permits greater flexibility in the treatment of geometrical irregularities and material variability and allows one to obtain values of dependent variables of their derivatives more directly at arbitrary points in the domain considered." These considerations illustrate the advantages of the finite element method, at least for mine water inflow in fractured rocks.

The finite element formulation ultimately results in a matrix that must be solved. From that point on it differs little from the finite difference solution that also requires solution of a matrix.

The so-called boundary integral method is also of recent origin. This method produces a solution for a field problem only in terms of the conditions at the boundary. In general, the integral method is used only where the flow field is homogeneous or it can be divided into a few homogeneous regions (Faust and Mercer, 1980). Two different methods are used, the indirect method and the direct method. In the indirect method the principle of superposition is used with a set of fictitious point sources along the boundary that, when added up, produce the given boundary conditions. In the direct method Green's theorem and matrix techniques are used to solve for the unknown heads or fluxes along the boundary. This process has the effect of reducing the number of dimensions in the problem by one. None of the codes that we have reviewed use the boundary integral method.

Regardless of the method used, the principal problem is the selection of the values of transmissivity and storage coefficient that should be incorporated into it. The most realistic solution to this problem is to use values of these coefficients that have been determined from multiple well pump tests in the area of interest. However, in many cases the results from a pump test are from such a small area in comparison to an entire flow system that the values are not representative of the entire aquifer. Consequently, the comparison between the model results and the field results is frequently poor. For this reason many "modelers" utilize a technique known as the "inverse" approach. That is, they use a mathematical model, either finite difference or finite element, but solve the matrix equations for the hydraulic conductivity or transmissivity and storage coefficient and match the model's input against measured hydraulic head. Values of transmissivity are obtained that give the same heads in wells in the model simulation as actually occur in the field.

When the process of calibration was developed originally, it was shown to be rather inaccurate. The problem results from the fact that when large, relatively horizontal aquifers are being investigated without the existence of pumping cones, transmissivity is highly dependent on the head measured. If the slope of the water table is relatively flat, the error in measuring the head in the well may be larger than the differences in head due to the gradient. Thus, the transmissivity can be in error and, in fact, it can be shown mathematically that the error in transmissivity may be unbounded (Cooley, 1979). Several recent papers present methods for determining transmissivity much more accurately (Frind and Pinder, 1974; Gillham and Farvolden, 1974; Neuman, 1980; Neuman et al., 1980a; Neuman and Yakowitz, 1979). The field data needed to implement the inverse problem approach are head measurements in authentic observation wells.

COMPUTER CODES

As mentioned previously, approximately 300 computer codes were reviewed before selecting one for use in this manuscript. The source of the majority of these codes is the Holcomb Research Institute at Butler University, Indianapolis, IN, that has published a catalog of computer programs obtained through surveys of researchers. Another source of information is a TRIS computer search of programs. This search located several unpublished Ph.D. dissertations in which computer programs have been developed.

One criterion that must be met for use of a model for water inflow prediction is that the model must be able to handle a seepage surface. The list of approximately 300 programs from the Holcomb Research Institute was reduced to three that would consider a seepage surface along with ten programs that can implement the inverse approach. Several articles and programs were reviewed that considered flow in fractures by techniques other than Darcy-type flow. Several of these considered flow through fractures and then by means of volume averaging develop essentially a Darcian-type equation. The conductivity in this case is not the value of the conductivity of the porous media itself but is dependent essentially on the location, direction, and size of fractures. We question whether the additional complication of such a fracture flow model is justifiable because the description of most fractured porous media (based on field data) is not sufficient to justify the additional details of such a model.

Ultimately we selected and decided to investigate more thoroughly four different computer programs. These are UNSAT2, FLUMP, TRUST, and WATGEN. The first two are finite element models, the third is an integrated finite difference model, and the fourth is a combination finite element and finite difference model. UNSAT2 was developed in Israel, FLUMP and TRUST were developed at the University of California, Berkeley, and WATGEN was developed at The Pennsylvania State University.

The general criteria used in the selection of these models and how well the models satisfy the criteria are presented in Table 4.

Several of the criteria shown in Table 4 require explanation. The first criterion, seepage surface, is the most important one for analysis of flow into an underground mine. A seepage surface boundary condition is one in which the pressure on the water is atmospheric and water comes from the aquifer at a particular elevation that reflects the total potential. This condition is necessary in order to produce simulated inflow into a mine realistically.

The second criterion, the characteristic of being able to simulate a perched water table, is necessary because in layered materials perched water tables develop above the less permeable hydrostratigraphic units as drainage occurs. Several ground water models will allow such perching even though they consider saturated flow only. This is done by linking the flow in a particular layer to the layer below the low permeability layer. The water moves from one permeable layer to the next permeable layer due to differences in hydraulic head. We are of the opinion that it is preferable to allow the perched water table to develop by simulating the unsaturated flow zone as it actually occurs in the field.

The third criterion, the consideration of nonhomogeneous materials, is necessary because in most field situations, nonhomogeneous materials occur. The same rationale applies to the fourth criterion, anisotropy.

Table 4. Criteria of selection

Characteristics	UNSAT2	FLUMP	TRUST	WATGEN
Necessary characteristics				
Seepage surface	yes	yes	yes	yes
Perched water table	yes	yes	yes	yes
Nonhomogeneous materials	yes	yes	yes	yes
Anisotropic materials	yes	yes	yes*	yes
Desirable characteristics				
Saturated-unsaturated flow	yes	yes	yes	yes
Irregular mesh	yes	yes	yes	no
Hysteresis	no	no	yes	yes
Consolidation	no	no	yes	no
Flow in fractures	no	no	no	yes†
Atmospheric-controlled evaporation and infiltration	yes	no	no	no

* Cannot conveniently handle arbitrary anisotropy.
† Is described as considering flow in fractures but uses essentially a Darcy equation.

The desirable characteristics which are listed include saturated and unsaturated flow. This capability, as stated previously, is highly desirable for the simulation of perched water tables. It is necessary also for simulation of infiltration or evaporation from the ground surface. It is desirable to have the option of a very irregular mesh because many hydrostratigraphic units in the vicinity of mine sites are very irregular in shape. Hysteresis is the characteristic of a program by which the porous medium has a different permeability when it is approaching saturation than when it is draining or becoming dryer. The ability to simulate hysteresis is desirable if infiltration from the ground surface is to be considered while drainage is occurring into the mine. Simulation of consolidation is desirable because often, as materials drain, some consolidation occurs, particularly in clayey materials. If infiltration and evaporation from the ground surface are to be considered, it is desirable also to use a code that allows for atmospheric control of evaporation. By this we mean that the maximum value of evaporation is controlled by the atmosphere but the actual rate of evaporation is controlled by the rate of upward movement. For infiltration, the maximum rate is controlled by the rate of rainfall while the actual rate is controlled by the rate of downward movement.

A detailed discussion of the four models follows, along with a discussion of the advantages and disadvantages of each.

UNSAT2

This program was obtained from Neuman (1974). UNSAT2 is a finite element based program that uses the Galerkin method leading to an integral representation of the problem. It will handle nonuniform flow regions having irregular boundaries and arbitrary degrees of local anisotropy. Flow may be saturated or unsaturated and can take place in a vertical plane, a horizontal plane, or a three-dimensional system with radial symmetry. Particular emphasis is placed on the simulation of atmospheric boundaries. This program also will treat flow to a fully or partially penetrating well of finite radius considering well bore storage.

The equation on which this program is based is:

$$\frac{\partial}{\partial X_i}\left[K^r(\psi)\,K_{ij}^s\,\frac{\partial\psi}{\partial X_j} + K^r(\psi)\,K_{i3}^s\right] - [C(\psi) + \beta S_s]\,\frac{\partial\psi}{\partial t} + S = 0 \qquad (70)$$

where X_i ($i = 1,2,3$) is the spatial coordinates (3 is vertical), K^r is the relative conductivity (a function of pressure head ψ), ψ is the pressure head, K_{ij}^s is the conductivity tensor at saturation, C is the specific moisture capacity $\partial\theta/\partial\psi$, β is 1 in the saturated zone and 0 in the unsaturated zone, S_s is the specific storage of the hydrostratigraphic unit under consideration, t is the time, and S is the positive sink term (or equivalently, a negative source term). S_s reflects the combined elastic properties of the porous medium and the water if lateral strains are negligible. S is the volume of water removed from the system by some negative source such as evaporation or evapotranspiration. Eq. 70 is essentially a combination of the continuity and Darcy equations with the source or sink term added.

The Galerkin method involves obtaining a solution for the pressures, ψ, at the node points and minimizing the residual between the solution value of ψ and the true value. The flow region is first discretized into a network of triangles and the pressure is determined at each node point. The time derivative is written in finite difference form and the resulting set of linear algebraic equations is solved by a Gaussian elimination algorithm. Since the original set of equations obtained from Eq. 70 is nonlinear, this solution must be improved by iteration. This iterative process is fully explicit and therefore is very rapid. However, the result is only an approximation so another iteration is necessary. If the maximum pressure change between the iterations is of a necessary small size, then the solution is complete for that time step. The time is then advanced by a specified increment and the iterative process proceeds as before. The initial time step is always specified and a time multiplier may be used to increase the time steps as the solution proceeds. This leads to one of the most difficult problems in using UNSAT2. If the initial time step is too large, a solution will not be reached and the iterative procedure stops. If the time multiplier is too large, the solution will advance for some time until the time step is too large and instabilities occur whereupon the iterative procedure stops.

Under the batch mode of operation as used ordinarily on large computer systems, this is an inefficient method of operation because it is not possible to determine whether the program will run except by trial and error. The program may run for a portion of the specified time and then stop. The only recourse is to begin again with a smaller time multiplier. With an interactive system such as that used by the authors, this shortcoming is relatively easy to correct. The solution can be monitored at the CRT terminal. If iteration stops, the final pressures can be replaced in the data input file and the program restarted from where it stopped rather than at the beginning.

UNSAT2 has been verified with laboratory data. We have verified it with manmade ponds constructed for oil shale wastes (Bloomsburg and Wells, 1978). It has also been used in several field problems but it is our opinion that in none of these cases were sufficient field data available for true verification.

The data input for UNSAT2 includes hydraulic conductivity and tables of values of moisture content vs. capillary pressure and moisture content vs. relative conductivity for each hydrostratigraphic unit in the flow field. The input data are coordinates and initial and boundary conditions for each node point as well as the node points and material type of each element. The authors use an interactive system consisting of the DEC PDP-11/23, wherein the coordinates of the node points are input with a digitizer. This method expedites the data input. The output of UNSAT2 consists of a table of pressures and rates of inflow or outflow at each node point. The time since starting, the cumulative inflow, and moisture content at unsaturated nodes are given also. UNSAT2 requires input of certain unsaturated flow parameters. These parameters are essential for the successful operation of the model even though the flow field may remain saturated throughout the simulation. Other models may be appropriate to the prediction of underground mine water inflow. These other models, especially the saturated flow models, may not require the unsaturated flow parameters required in UNSAT2. A description of unsaturated flow and its essential parameters is included as Appendix E.

FLUMP

FLUMP is a finite element program developed at the University of California (Narasimhan, et al., 1978). It is based basically on the same equation as UNSAT2. This equation is essentially a Richard-type equation that results from substituting the Darcy equation into the unsteady flow continuity equation. The matrix equation that evolves from the Richard equation is

$$Ah + Dh' = Q$$

where A is a conductance matrix, D is the capacity matrix, h is hydraulic head, h' is the time derivative, and Q is a vector representing sources or sinks.

According to Narasimhan (1978), trouble has been encountered in the program's ability to solve this equation because the capacity matrix D is non-diagonal. This introduces a problem of having a mass balance in the vicinity of each node not completely satisfied although the overall mass balance in the entire finite element grid is preserved. It has been shown also that if the matrix is diagonal, the permissible value of time steps is larger and leads to a realistic solution.

Neuman (1977) describes an alternate form of the capacity matrix that has many advantages over the ordinary approach. In this approach the system of differential equations expressed by the foregoing equation is discretized in time by finite differences. This procedure makes it possible to treat the equations either explicitly, implicitly, or by a combination of both schemes. The implicit equations are solved by an iterative procedure rather than by Gaussian elimination. In the same paper, the authors derived stability and convergence criteria for the explicit-implicit scheme as well as convergent criteria for the point iterative technique.

Neuman (1977) also demonstrates that the explicit-implicit scheme converges to the exact solution of the equation at each node at which stability and local symmetry conditions are satisfied. This local nature of the stability criteria together with the use of the point iterative technique introduces the possibility of solving the finite element equations explicitly at any nodes at which the stability criteria are satisfied and implicitly at other nodes during any particular time step. A final correction is made to the explicitly calculated Δh values when required to maintain conservation of mass. This approach is called the mixed explicit-implicit solution method. The method was used originally in a computer program called TRUMP that was designed for heat flow by conduction, convection, and radiation. FLUMP is a modification of TRUMP designed specifically for fluid flow in porous media. Another advantage of FLUMP is that the size of the time step is controlled automatically. The first time step is always 10^{-12}; it is then increased by a time multiplier. If the time step gets too large, so that instabilities occur, the time step is reduced automatically. At such time as the solution again becomes stable, the time step is again increased. The printed output from the program gives the pressure at each node, the amount of fluid contained in the subdomain of each node, change in the amount of fluid in each subdomain during a particular time increment as well as from the start of the problem. The total fluid content in the system, the flux across the boundary of the system, and the flux into or out of a subdomain of each node are printed out also. This information makes it possible to check continuously the material balance at a node. Due to the method of solution that is used, the numbering of nodes and elements is completely arbitrary whereas in many other programs (UNSAT2), it is necessary to number the nodes in a particular order to decrease the band width of the conductance matrix. FLUMP does not have some of the options of UNSAT2, such as atmospheric-controlled evaporation or infiltration, or the option of considering water used by plants. These are not considered to be mandatory requirements for the analysis of mine water inflow, but for some problems they might be desirable. A number of problems are reported for verification of FLUMP, including buildup of a ground water ridge under free surface conditions, infiltration into an unsaturated soil column, drainage from a saturated/unsaturated system, and confined axisymmetric flow to a well with well bore storage. These problems all show reasonably good agreement with other numerical solutions, analytical solutions, or experimental data. The program may be used in a vertical plane, in a horizontal plane, or in an axisymmetric problem.

TRUST

TRUST is a program presented in a series of three articles (Neuman, 1977, 1978a, 1978b). These papers cover, respectively, the theory, the algorithm, and the applications. TRUST is a general three-dimensional model which will simulate ground water flow in variably saturated, deformable, heterogeneous porous media. It considers consolidation in one dimension only, namely the vertical dimension. The deformation of the soil skeleton in this program is handled according to Terzaghi's one-dimensional consolidation theory. The program also will consider hysteresis that has been described previously.

The integrated finite difference method of descretization is used in TRUST. This method differs from the finite element method in the way that an element is described because the area surrounding the node point is attached to that node point. The area for a node point is the area that is closer to that node than to any other node. This concept is shown in Fig. 58. Although this method is somewhat different from the finite element method, the method of solving the set of equations is the same as that used in FLUMP. This method is the mixed explicit-implicit iterative scheme. The main advantage of TRUST is that it attempts to pose and solve a problem physically. This procedure is much more straightforward than that used in finite element programs. The chief limitation of TRUST is that the finite difference gradient approximation is not able to handle general tensorial properties such as anisotropy and stress. The anisotropy problem can be overcome somewhat by orienting the elements parallel to the principal axis of anisotropy. This is not necessary in a finite element model.

Numerically the strength of TRUST lies in the fact that it is multidimensional and the iterative scheme used avoids the large requirements of computer storage associated with direct solution algorithms. The problems that have been used for verification of this program include a one-dimensional problem of saturated flow with consolidation. This verification was first done on a laboratory column of clay, then on a column of clay slurry, and finally in a field situation near Pixley, CA. The verification for unsaturated flow includes infiltra-

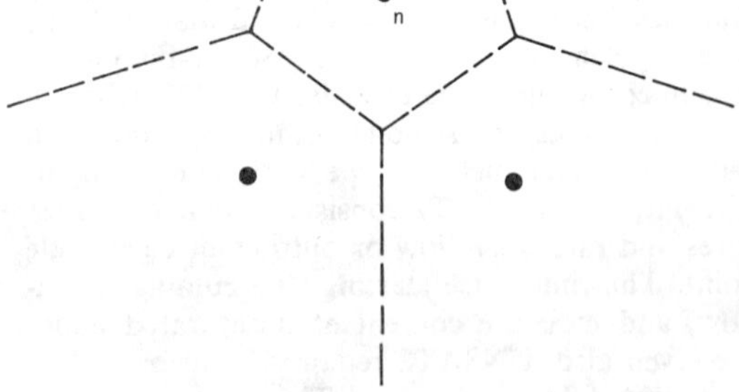

Fig. 58. Node system for the integrated finite difference method.

tion into a moderately saturated soil, infiltration into a column of extremely dry soil, axisymmetric flow to a soil/water sampler, drainage from a one-dimensional soil column, drainage from a sandbox (this particular problem was also used as verification for FLUMP), consolidation around an excavation in soft clay, and pore pressure generation and dissipation.

The data input for TRUST is organized into the following four parts:

Part 1: Pretitle and title cards.

Part 2: Block cards.

Part 3: Problem continuation card or data end card.

Part 4: Problem termination card.

These terms are self-explanatory except for the block cards which are described as follows.

Block 1. The problems, controls, limits, and constants: These data include the number of time steps between data outputs, identification numbers of nodes for which potential, rate of change of potential, source rate and time will be written for each time step. The option of having a deck of cards punched with the final results, in case it is desired to continue or restart the program is available also. Several of the cards in this block concern the time steps and the time of running.

Block 2. System and material properties: The system properties include hysteresis, specific gravity of the saturated soil, and thickness of the flow region. The material properties consist of tabulated values of degree of saturation and capillary pressure and tabulated values of permeability as a function of pressure. The coefficient of compressibility that is used for one-dimensional consolidation calculations also is placed in this block as is the permeability, specific storage, swelling index, and compression index.

Block 3. Properties of the fluid: Fluid properties are viscosity, compressibility, and density.

Block 4. The node description: The program will generate a specified number of identical nodes, but one card must exist for each specified node giving the material number for that node and the coordinates of the node.

Block 5. Internal fluid flow connections: This gives the identification numbers and descriptions of connected nodes.

Block 6. External fluid flow connections: A card exists for each boundary node and surface node that includes the numbers and surface fluid transfer coefficients.

Block 7. Boundary nodes and external potentials: A card for each boundary node gives the node number and external or boundary potentials. An option of having a sinusoidal variation of boundary potential is available also.

Block 8. Boundary nodes and variable sources or sinks: These cards contain the node identification number, a fluid generation rate, and tabulated values of fluid generation.

Block 9. Initial conditions: These cards contain node identification numbers, initial potential, fluid generation rate, and the initial preconsolidation stress.

Block 10. Dimensionless variables: These cards contain identification numbers of the nodes for which dimensionless quantities are to be calculated.

Several of the blocks previously described consist of only one card or in the case of node points, a card for each node point.

A copy of TRUST on tape was obtained by the authors from R. W. Nelson of Battelle Pacific Northwest Laboratory at Richland, WA. According to Nelson, TRUST is not easy to operate but is a very powerful program and is well worth utilizing. The documentation that is available for the program is somewhat incomplete.

WATGEN

This program was developed at The Pennsylvania State University through a contract with the US Bureau of Mines. The objective of this project was to develop a model that would consider a number of aspects of flow in coal mines including airflow, flow of acid water, and flow of methane. The information presented herein is derived from Manula (1975).

On first glance this program appears to be very well documented and would be reasonably easy to use. However, a detailed examination of the report shows that the approach is weak in the area of unsaturated flow. The terms permeability and hydraulic conductivity are used interchangeably as are the terms relative and effective permeability. Inconsistent systems of units are used throughout the program so that equations that are solved include several numerical values that would not be included if consistent units were used. The program is said to be a combination of finite difference and finite element methods. However, it is considerably more restrictive than most finite element programs in that the region of solution must be a very regular shape. The mesh is not constructed so as to correspond to the boundaries between different types of material. In fact, the mesh is completely independent of where the materials change from one type to another. If a particular block in the mesh has several different materials in it, then a weighted average is used for all properties for that block.

The meaning of the combination finite difference and finite element method is that the equations are written in finite difference form. The stated advantages for this approach are that the mathematics are simpler; the program is compatible with machine oriented methods of data storage and retrieval; the method is versatile in solving nonlinear boundary value problems; and computer time can be optimized by using appropriate solution procedures. The use of the finite element analysis is present simply in the concept of piecewise approximations of the media properties. This makes possible the inclusion of natural boundary conditions in the formulation; the size of the elements can be varied and nonhomogeneity and anisotropy can be handled.

The data input for the program involves a great number of values but is relatively simple because it has a free format. The first few cards are problem controls, program controls, problem parameters, capillary pressure, water relative permeability, air permeability, hysteresis data, meteorological data, geologic and fluid data, and general data. One inconvenient factor is that a card is used for each block in the mesh and values of all properties of that block such as relative permeability, saturation, and conductivity must be specified for each block. In most programs this is accomplished by specifying a material number for a particular element; the material properties are then listed for that one material. The method used in this program involves inserting a great deal more data than is required by the other programs.

WATGEN was run on a mine in Pennsylvania for verification. The verification was accomplished by comparing the actual mine drainage with that simulated by the program. The comparison was very good with a maximum difference of about 28%. This relatively good comparison is somewhat surprising considering that the data used for input were obtained not in the field but from the literature. The input data included the initial position of the water table, initial soil water saturation, elastic constants, fracture permeability, residual water saturation, saturated soil permeability, and the storage coefficient. The program was run once with the initial data that were obtained from the literature; the results were as stated.

Other Programs

We have examined two other programs in detail. They are presented in O'Neill (1977) and Sawyer (1978). Both these programs consider the flow of both liquid and heat in fractured porous media. In the consideration of the fractured characteristics a fracture permeability and the pore permeability are used. The addition of these terms produces the permeability tensor that is used in the solution of the problem. The mesh used in this three-dimensional program must be rectangular in shape. The position of the nodes are specified and the elements are then generated automatically. It appears that this program would be difficult to fit to a region with nonregular material boundaries such as are likely to be present in the vicinity of a mine.

The program by Sawyer (1978) is called SIMPACK. It is a diffusivity model for fluid flow and heat conduction in porous media. It is three-dimensional and the formulation is by finite difference methods. It is essentially limited to a rectangular mesh, making it difficult to simulate natural materials that would occur near a mine. This program has been used by at least one mining company but only after extensive modification of the program. Neither of these two programs have had any field verification.

6

Developing a Mine Water Inflow Method

The purpose of this section is to demonstrate the development of a method that we believe can be used for simulating and predicting mine water inflow. It will also show how important it is for personnel who model mine water inflow to have considerable experience with this subject, and with the conversion of hydrogeologic frameworks to conceptual models.

Several of the models described previously are said to be three dimensional. However, closer examination reveals that there is considerable limitation in the variation possible in the third dimension. A true three-dimensional analysis would require a very large program. Therefore, it seems more reasonable to develop a procedure by which information from a two-dimensional model can be used for analysis of a three-dimensional problem. For a vertical mine shaft this is relatively easy because most two-dimensional models have an option for axisymmetric flow. If it can be assumed that axisymmetry exists about the centerline of the shaft it can be modeled easily as has been illustrated previously.

For a horizontal drift or tunnel the analysis is not nearly as straightforward. However, the flow can be divided into two parts as indicated in Fig. 59. Part 1 considers the flow into the side of the drift while part 2 considers the flow into the end of the drift. To calculate the flow rate into the entire drift from the unsteady flow into a unit length the instantaneous flow per unit length may be integrated over the total length of the shaft. This procedure is shown in Fig. 60. The radial flow through any incremental length of the shaft is q. The total flow rate, Q, into the drift at any particular time is given by:

$$Q = \int_{o}^{L} q \, dx. \qquad (71)$$

The discharge q is a function of the time during which a particular segment has been draining that is dependent on the location x. In other words, a location at $x = 0$ has been draining during the entire time of expansion while a segment at $x = L$ has only started to drain. Therefore, the time of draining at any point, x, is given by:

$$t = \frac{L - x}{L} t' \qquad (72)$$

where t' is the time over which the mine has been expanded. To determine the relationship between q and t it is necessary to refer to Polubarinova-Kochina (1962, pp. 566–569) and the equation given for the unsteady flow problem. This is:

$$q = q_0 + q_1 t + q_2 t^2 + \ldots \qquad (73)$$

where q is the steady flow, q_1, q_2 are the coefficients, and t is the time since drainage began. The expressions for q_0, q_1, and q_2 are given as:

$$q_0 = \frac{2\pi KH}{\ln \dfrac{2H}{r}}$$

$$q_1 = \frac{q_0 (3q_0 - 4K\pi H)}{4n\pi H^2 \ln \dfrac{2H}{r}}$$

$$q_2 = \frac{q_1 (9q_0 - 4K\pi H)}{8n\pi H^2 \ln \dfrac{2H}{r}} + q_0^2 \frac{(27q_0 - 28K\pi H)}{8(2n\pi)^2 H^4 \ln \dfrac{2H}{r}} \qquad (74)$$

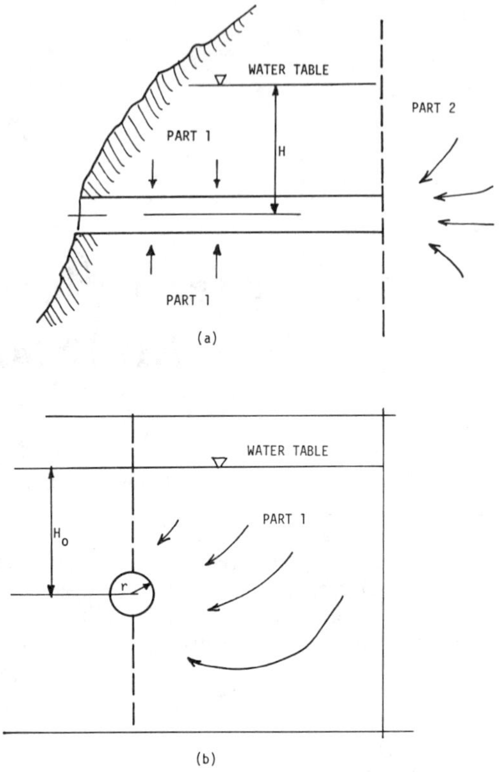

Fig. 59. Regions for flow into a mine drift.

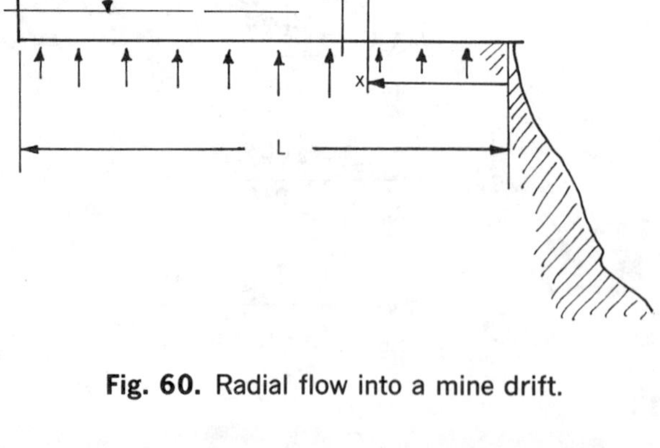

Fig. 60. Radial flow into a mine drift.

$$+ \frac{(K\pi)^2 q_0}{(2n\pi)^2 H^2 \ln \dfrac{2H}{r}}$$

where K is the hydraulic conductivity, H is the height of water table above the center of the drift, r is the radius of the drift, and n is the drainable porosity.

Dividing Eq. 73 by q_0 yields the following expression:

$$\frac{q}{q_0} = 1 + \frac{q_1}{q_0} t + \frac{q_2}{q_0} t^2 + \ldots$$

Substituting the coefficients q_0, q_1, and q_2, rearranging, and simplifying yields:

$$\frac{q}{q_0} = \frac{1.5 - \ln \dfrac{2H}{r}}{\left(\ln \dfrac{2H}{r}\right)^2} \left(\frac{Kt}{nH}\right) + \frac{13.5 + 1.5 \ln \dfrac{2H}{r} - 5\left(\ln \dfrac{2H}{r}\right)^2 + \left(\ln \dfrac{2H}{r}\right)^3}{4\left(\ln \dfrac{2H}{r}\right)^4} \left(\frac{Kt}{nH}\right)^2 + \ldots \qquad (75)$$

This is an equation giving dimensionless discharge in terms of dimensionless time Kt/nH with $2H/r$ as a parameter. The variable Kt/nH has been used in other unsteady drainage problems (Brown, et al., 1977; Glover, 1966).

Eq. 75 may be written as

$$\frac{q}{q_0} = 1 + a\left(\frac{Kt}{nH}\right) + b\left(\frac{Kt}{nH}\right)^2 + \ldots \qquad (76)$$

Fig. 61 shows plots of this equation for two values of $2H/r$. In the development of Eq. 76 there is no definition of what is meant by "small" values of time. Therefore, we have simulated this problem with UNSAT2 to evaluate the equations.

For simulation the finite element meshes shown in Figs. 62 and 63 were used for $2H/r = 5$ and $2H/r = 20$, respectively. Fig. 63 appears not to show a circular drift. This occurs because the horizontal scale is different than the vertical scale. The coordinates of the node points when the problem was run did produce a circular drift.

A hypothetical material with a hydraulic conductivity of 1 ft (0.3 m) per day, porosity of 0.2, and specific storage of 10^{-6} was used with the meshes. The curves defining the capillary pressure head vs. moisture content and relative conductivity vs. moisture content relationships are shown in Fig.

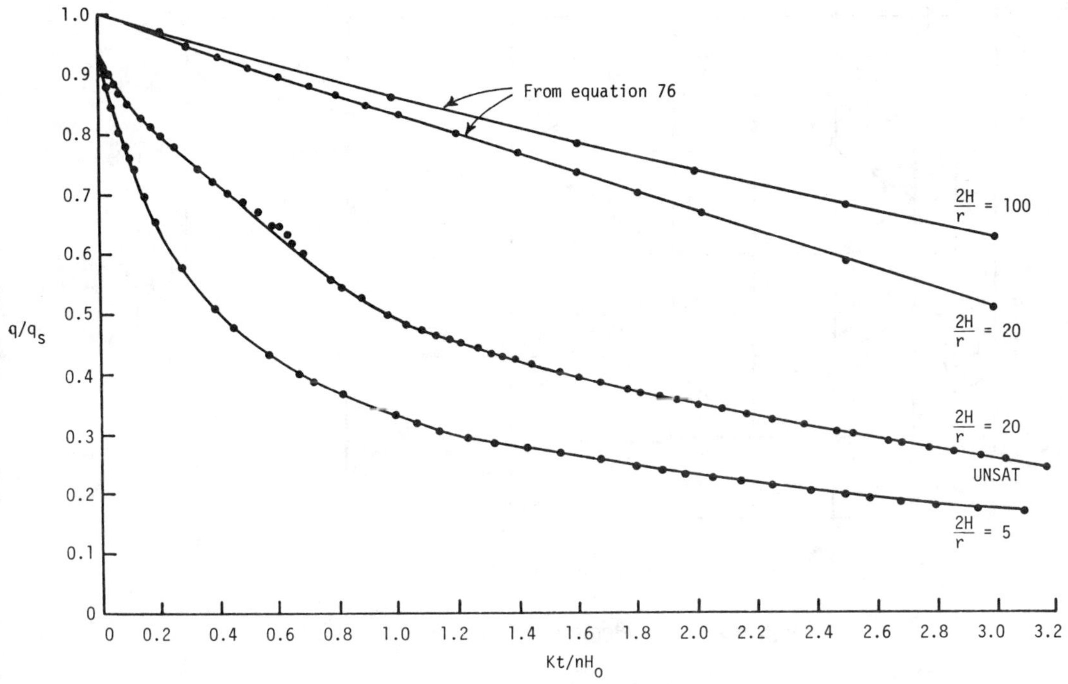

Fig. 61. Inflow rate to the drift.

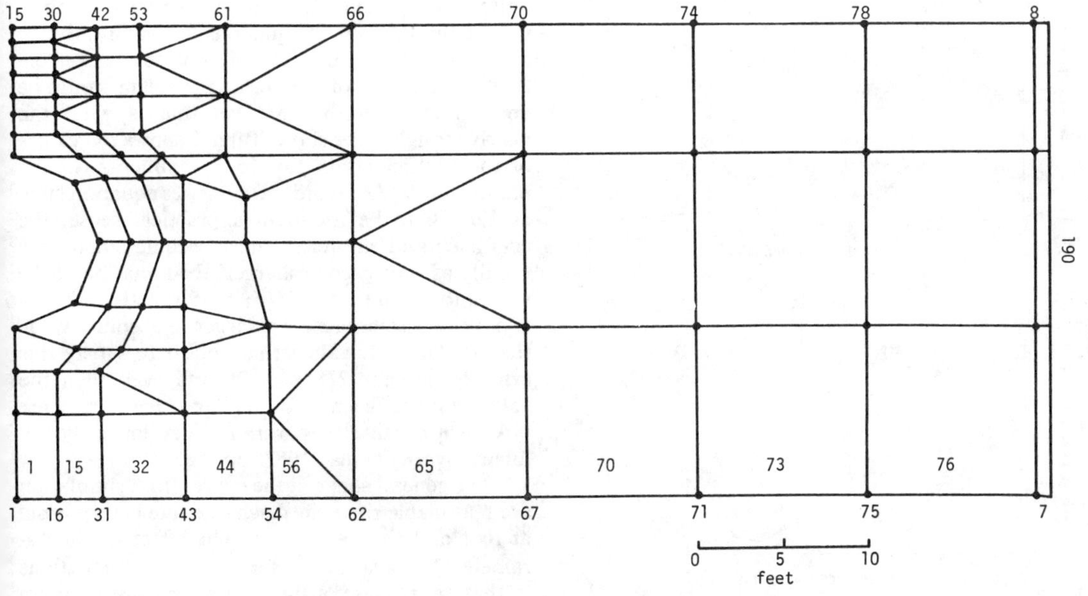

Fig. 62. Finite element mesh for $2H/r = 5$.

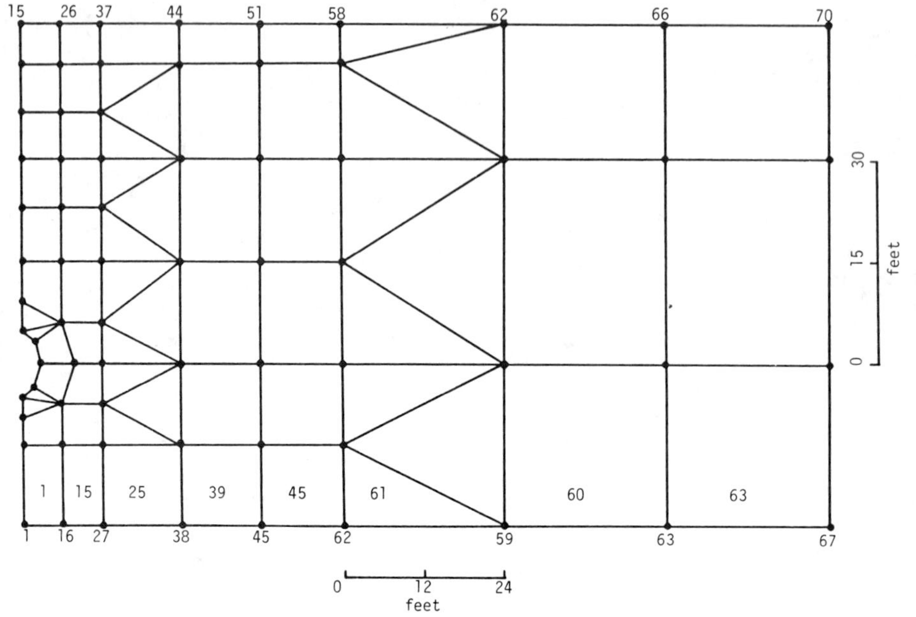

Fig. 63. Finite element mesh for $2H/r = 20$.

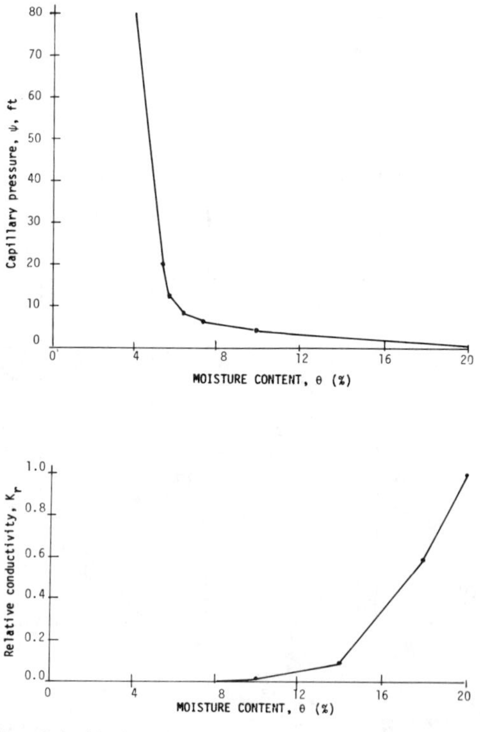

Fig. 64. Hydraulic properties for hypothetical material, drift inflow.

64. Although these are hypothetical curves they are reasonable for a coarse sandstone.

Returning to Fig. 61 it is obvious there is little agreement between the theoretical curves and the curves from simulation. Several plausible explanations are possible, but the most probable cause is that the theoretical equations are valid only for an infinitesimal time. It is simply not reasonable that the decrease of flow in this problem would be approximately linear, and yet that is what the nearly straight lines show. Table 5 shows the values of the coefficients in Eq. 76 as a function of the parameter $2H/r$. In order for a series equation such as Eq. 76 to be useful in a practical sense, the coefficients of the terms must become small very rapidly as time becomes larger. Examination of the values for a and b for $2H/r = 5$ reveals that this is not the case because b is larger in absolute value than a. The only values that appear to satisfy this criterion are for $2H/r = 20$ and even then the values of coefficients for higher order terms are unknown because they were not developed by Polubarinova-Kochina (1962, pp. 565–570).

The general shape of the curves from simulation are reasonable since the discharge rate is very rapid at first and then levels out. The effect of the parameter $2H/r$ is similar for both sets of equations in that an increase of the parameter causes a more gradual decrease of discharge.

Although it does not appear that Eq. 75 is use-

Table 5. Coefficients for Equation 74

$\frac{2H}{r}$	H, ft	a	b	Eq. 74 q_o, gpm	UNSAT2 q_o, gpm
5	12.5	−0.042249	0.265724	0.254	0.258
10	25	−0.151377	0.023590	0.354	
20	50	−0.166666	0.000020	0.545	0.554
50	125	−0.157608	0.002901	1.043	
100	250	−0.146418	0.006689	1.772	

ful, Eq. 74 for the initial rate of flow is of value. Table 5 also shows a comparison of the initial rate of discharge from Eq. 74 and from UNSAT2. The agreement is very good.

Figs. 65 and 66 show the locations of the water table at various values of Kt/nH for the drifts where $2H/r = 20$ and 5, respectively. The shapes of these curves are as would be expected with the steepest gradients near the drift.

Fig. 67 shows the ratio of the water table height h over the center of the drift to the original water height H as a function of the time parameter Kt/nH for the two drifts. This figure illustrates one of the problems with any type of numerical model. The abrupt drops in the water tables shown by the dashed lines are the result of nodes 8 and 9 on the surface of the drift desaturating and causing an abrupt change of the pressure in surrounding nodes. To keep the size of the finite element mesh relatively small only five nodes were used for the surface of the drift. Thus the nodes are spaced relatively widely and the discrete jumps occur. These would be reduced by using a finer mesh. In the previous figures there appeared to be no problem associated with these relatively coarse meshes.

In all numerical models the difference between the solution obtained and the correct value is due to truncation, roundoff, errors in the model, and errors in programming. We have used UNSAT2 sufficiently to ascertain the absence of programming errors and the limitations of the model used are understood. Therefore, the primary source of errors in this case is due to truncation (using too coarse a mesh) or roundoff. Roundoff errors are related to the number of algebraic operations yielding a particular solution; as long as the time steps are not too small these should be acceptable. The only way to evaluate the magnitude of truncation error is by comparison with a solution from some other method or by running the program with another mesh to see whether the new mesh produces a change in the solution. This is the procedure that would be followed with the data in Figs. 61, 65, 66, and 67 before the results could be used in any type of design problem.

Returning to the discussion of how Fig. 61 can be used for determining mine water inflow rate, q can be written as:

$$q = q_o f\left(\frac{Kt}{nH}, \frac{2H}{r}\right) \quad (77)$$

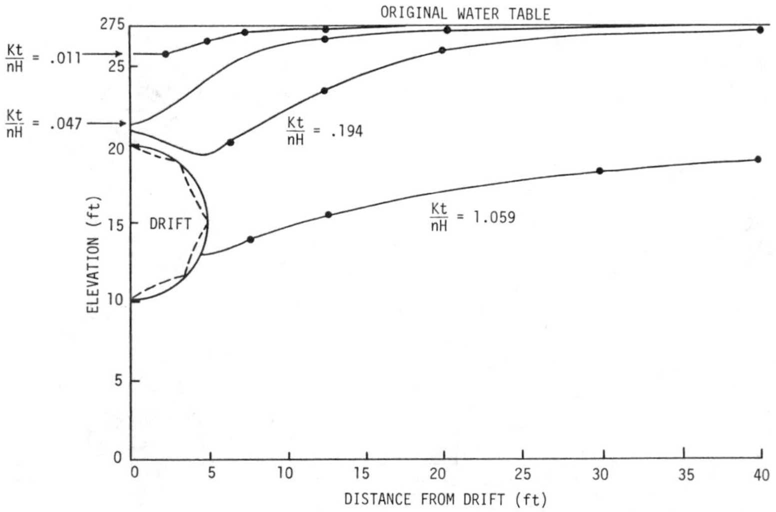

Fig. 65. Water table locations for drift $2H/r = 20$.

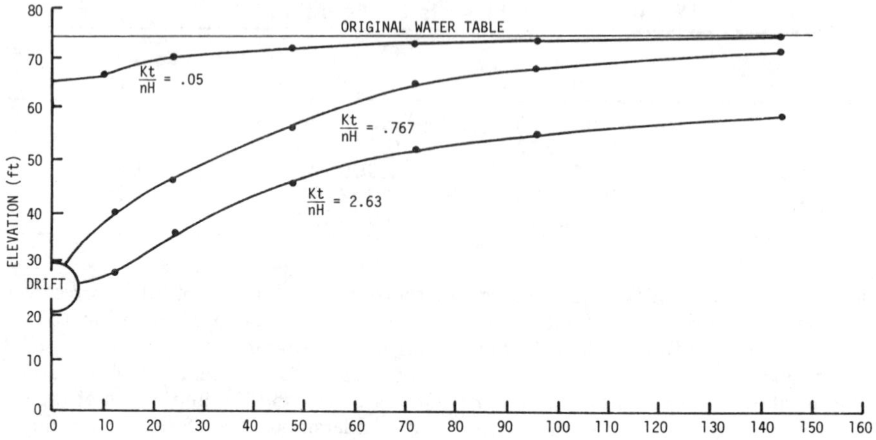

Fig. 66. Water table locations for drift $2H/r = 5$.

where the functional relationship is that given by Fig. 61. Eq. 72 is an expression for t so by substitution:

$$q = q_0 f\left(\frac{K\left(\frac{L-x}{L}\right)t'}{nH}, \frac{2H}{r}\right). \quad (78)$$

Substituting this expression into Eq. 71 gives

$$Q = q_0 \int_0^L f\left(\frac{K\left(\frac{L-x}{L}\right)t'}{nH}, \frac{2H}{r}\right) dx. \quad (79)$$

This equation could be evaluated for a particular time of expansion t' numerically or by approximating the function for q with a polynomial and integrating in closed form.

For the other part of the drift (the advancing face), there is no known analytical solution for either steady or unsteady conditions. A solution is possible if there is a boundary condition of constant potential at a constant radial distance away from the face, but this is not the case for a constant or varying water table. This problem can be run easily with the same finite element mesh as used previously by specifying an axisymmetric flow field. We simulated this condition for a short time for $2H/r = 5$, whereupon the initial flow into the end

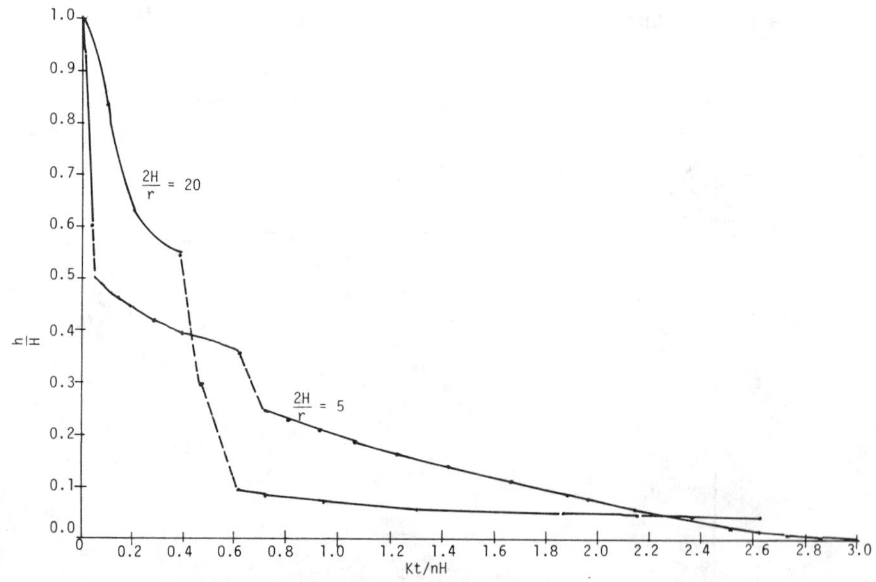

Fig. 67. Water table height above centerline of drift.

of the drift proved to be 0.413 gpm (0.0016 m^3/min). This rate is approximately the same as the flow into 2 lineal ft (0.6 m) of the shaft; therefore it is insignificant for a shaft that is of any appreciable length. Consequently, we chose to neglect it.

In order for the foregoing analysis to be used in an existing mine, it would be necessary to develop the curves in Fig. 61 for that mine. Nonhomogeneous or anisotropic conditions can be considered as well as any geometric shape of flow field. The most critical assumption used in this analysis is that the flow field is perpendicular to the axis of the drift. This condition is satisfied if the drift is straight but it is not satisfied if bends occur in the drift. Another assumption is that the two-dimensional flow field is to a large extent perpendicular to the axis of the drift. It would be possible to simulate a recharge constant potential or no-flow boundary at a specified distance from the drift. This would be difficult, however, if the drift was not parallel to the boundary. We have concluded that this method is one way of applying a two-dimensional model to a three-dimensional mine water inflow problem.

DISTRICT HYDROLOGY

The impact of mine dewatering on the ground and surface water systems and the use of mathematical models in predicting this impact will be considered here in general terms. The specific impact of each mine simulated will be considered in the section describing the simulation of that particular mine.

In general, mine dewatering will have the greatest effect on district hydrology if the mine is fairly shallow and the aquifer being dewatered is directly connected hydraulically to the surface water system. Mine dewatering would increase pumping lifts or dry up other wells in the aquifer and dry up springs if the water table dropped below the elevation of the discharge region of the aquifer being dewatered. Surface discharge could also be affected by changing gaining reaches of streams to losing reaches. Mine dewatering even in deep aquifers could cause an increase of pumping lifts in any wells producing from the same aquifers.

Water levels in an aquifer could rise if mine water pumped from another aquifer is reinjected into the first aquifer or if water is pumped from outside the cone of depression of mine dewatering and returned to the cone of depression of mine dewatering. Either of these situations may be necessary if it is not permissible to discharge the pumpage into surface streams.

Surface discharge may not be permitted if it would cause a significant degradation of water quality in the surface streams. Change of water quality may be due to a difference in dissolved chemicals or a difference in temperature between the pumped aquifer water and the receiving stream.

Predicting district impact of mine dewatering may require the use of several mathematical models because no completely three-dimensional models currently available will consider transport of dissolved constituents and/or heat. Therefore in any field situation it is necessary to reduce the problem to several simpler problems that may be modeled with existing programs. As an example, a water table aquifer that has direct hydraulic connection to the stream system in the region could often be modeled as a two-dimensional horizontal aquifer of uniform thickness. Impermeable boundaries, constant potential boundaries, and water removal at drainage wells could be considered and relationships between pumping rates, water table elevations, and stream channel inflow and outflow could be determined. UNSAT2 as well as several other ground water flow models will handle this type of simulation.

The flow of chemical constituents or heat could be predicted only by using a transport model. These models usually use as input a velocity field developed from a typical water flow model.

Mine dewatering from deeper aquifers not connected hydraulically to the surface water system could also be modeled with a two-dimensional horizontal model as long as the aquifer thickness is relatively uniform. UNSAT2 can be used in this type of problem but it will not consider leakage into or out of formations above or below the modeled formation. Other models are available that will consider such leakage. A significant advantage of UNSAT2, however, is that the boundaries may be changed with time to simulate the mined-out portion of the mine. Fault zones or any other type of boundary also may be modeled in this mode.

The most essential information that must be obtained for use in a district impact study would be the projected pumping rate during the life of the mine. This water must be discharged into either surface streams or injected into the ground at some other location. Whatever the disposition of the pumpage, given the proper field data collection program as discussed previously, the rate of flow can be predicted, from which the district impact can be determined.

In developing deep mines usually it is necessary to develop a vertical shaft that may be 15 to 20 ft (4.6 to 6.1 m) in diameter. The water inflow during development must be pumped and discharged into surface streams or reinjected into the ground. In order to predict the impact on the district hydrology it is necessary to know the quantity and quality of pumped water and the spatial relationship of the discharge-recharge areas.

As will be shown later during the simulation of an existing mine, UNSAT2 may be used in a unique manner to determine the inflow to the shaft continuously as the shaft develops. The uniqueness consists of operating the model in an axisymmetric mode with the shaft being the vertical axis of symmetry. As the shaft progresses downward the axis nodes are moved out to the radius of the shaft. Inflow may be allowed to occur at any point that the shaft will be open. After the rate of pumpage is determined the model facilitates the selection of the method of disposal that will give the most beneficial or least detrimental effect on district hydrology that can be achieved.

7

Existing Mines—Two Case Histories

INTRODUCTION

This manuscript is based on site inspections of five existing mines and the detailed analysis of two of the five. The detailed analysis and mathematical models of the two mines are presented herein in order to demonstrate a method of predicting mine water inflow using existing mines as a data base. Water management procedures were studied at all mines. Our original criteria for mine selection were that the mines be less than five years old and have an inflow rate in excess of 2000 gpm (7.6 m^3/min). Background hydrologic and hydrogeologic data were required so that two of the mines could be used for our computer modeling experiments.

Our original criteria proved to be difficult to meet. We contacted numerous mines via various sources. Many mines were unable to meet the criteria stated or were not interested in cooperating with us. Therefore, we allowed some flexibility in the application of these criteria to satisfy our original requirements.

Although the visitations at all the mines are not discussed herein the characteristics of the two mines simulated are presented. These simulated mines consist of a uranium mine in New Mexico and a lead mine in Missouri. The mines and owners will remain anonymous by request.

URANIUM MINE—CASE HISTORY NUMBER 1

Mine Location

The uranium mine that we simulated is located in northwestern New Mexico. The ground surface at the uranium mine is at an elevation of approximately 7000 ft (2134 m) above mean sea level. The climate is arid to semi-arid with an average annual precipitation of approximately 10 in. (25.4 cm) (John and West, 1963).

Geologic Setting

It is essential that a conceptual model of the stratigraphy, hydrostratigraphy, and structural geology of a mine site be developed prior to simulation by a mathematical model. If such a conceptual model is not developed prior to simulation major errors are difficult to avoid.

The uranium mine we have simulated is located in the San Juan Basin (Fig. 68), which is a structural basin located in northwestern New Mexico and adjacent parts of Colorado, Arizona, and Utah. The sedimentary rocks present in the deepest part of the basin are nearly 15,000 ft (4572 m) thick (John and West, 1963). The regional dip of the sedimentary rocks in the Grants area is generally northward toward the San Juan Basin; the regional attitude is modified locally by normal faults and folds (Hilbert, 1963). The rocks of interest in this area are sedimentary in origin and range in age from Pennsylvanian to Cretaceous. They rest on the Precambrian core of the Zuni Uplift. Associated intrusive and extrusive rocks are of Tertiary and Quaternary ages (Hilbert, 1963).

The deepest sedimentary unit of interest at this mine site is the Morrison Formation as described by Hilbert (1963). The Morrison Formation consists of three members in the general area of the mine site; the deepest member is the Recapture. The Recapture Member consists of alternating beds of gray sandstone and grayish red siltstone or mudstone with strata up to several feet in thickness. Only a few small deposits of uranium are contained within the Recapture Member. The Westwater Canyon Member overlies the Recapture Member of the Morrison Formation. The Westwater Canyon Member is approximately 150 ft (46 m) thick and consists of light brown to gray, poorly sorted, arkosic sandstone, and some interbedded gray mudstone. The Poison Canyon Formation denoted in subsequent discussions, tables, and figures was determined to be a part of the Westwater Canyon Member. The separate connotation is used for the description of mining operations and hydraulic properties but a separate physical description is not included. This member of the Morrison Formation contains many large deposits of uranium. This is

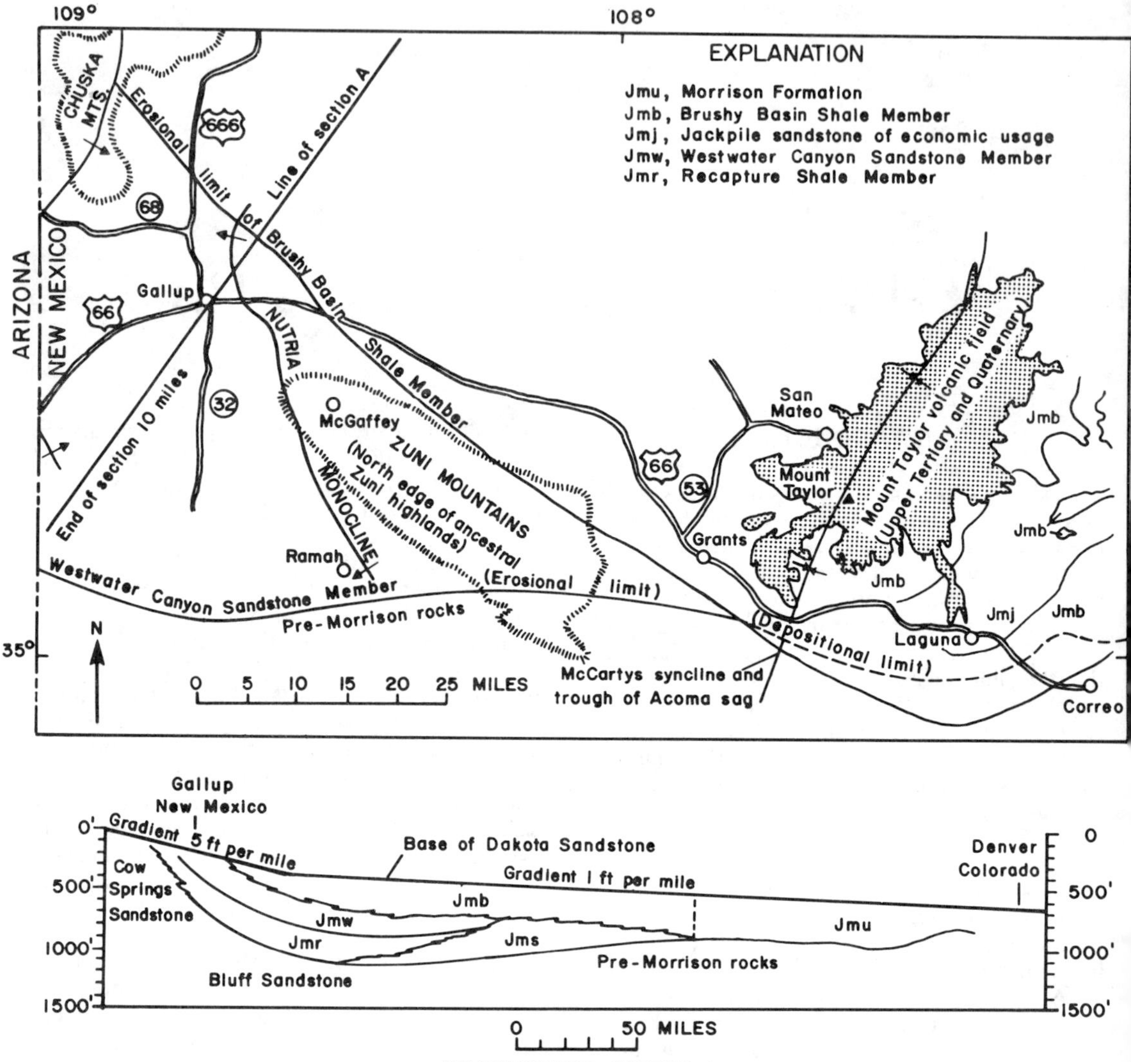

Fig. 68. Location map and geologic section (modified after Granger, 1968, Fig. 1).

the deepest member of the sedimentary sequence that will be mined. The Brushy Basin Member of the Morrison Formation overlies the Westwater Canyon Member and consists of greenish gray mudstone and thick arkosic sandstone units. Many deposits of uranium occur in the sandstone lenses of the Brushy Basin Member. The Dakota Sandstone overlies the Morrison Formation; it consists of a tan to gray, medium grained quartz sandstone with some interbedded carbonaceous shale and local coal lenses. The combined thickness of the Dakota Sandstone and the Brushy Basin Member is approximately 83 ft (25.3 m).

The Mancos shale in the Mesa Verde Group that overlies the previously described formations at this mine represents an intertonguing of the other formations due to alternating marine transgressions and regressions (Moench and Schlee, 1967). The three units of Mancos shale that overlie the previously described units are referred to as Lower, Middle, and Upper Tres Hermanos. These units are prominently sandstone layers that occupy the lower stratigraphic position in the Mancos shale. Moench (1967) described these units for the uranium mining area slightly east of Grants. The Lower Tres Hermanos typically is thinly bedded to massive with a few thin beds of siltstone and shale. It is fine to medium grained and fairly well sorted; it is approximately 118 ft (36 m) thick. The Middle Tres Hermanos typically is thinly to thickly bedded.

The coarsest material occurs near the top of the unit; it is fine to medium grained, well sorted, and fairly well cemented. This unit is approximately 70 ft (21 m) thick.

The Gallup sandstone of the Mesa Verde Group overlies the Mancos shale members described previously. This is the lowest unit of the Mesa Verde Group; it is fine grained, moderately well sorted, and poorly to well cemented with silica. It is approximately 270 ft (82 m) thick.

The Dilco Coal Member of the Crevice Canyon Formation overlies the Gallup Sandstone; it consists of interbedded sandstone and shale or siltstone. The sandstone is fine to very fine grained, firmly calcite cemented, and flat bedded. The siltstone is laminated to massive and friable to well cemented with calcite. The shale is laminated and fissile. This unit is approximately 80 ft (24 m) thick.

The Mulatto Tongue Member of the Mancos Shale overlies the Dilco Member; it includes a few sandstone strata. The graded shale sandstone sequence consists of shale or siltstone and very fine grained sandstone that may or may not be well cemented. The Mulatto Tongue Member is approximately 57 ft (17 m) thick.

The Dalton Sandstone Member of the Crevice Canyon Formation overlies the Mulatto Tongue Member. This member is fine grained and well cemented near its base; it grades downward into a less well cemented very fine grained sandstone in the upper part of the Mulatto Tongue Member. The top is fine to medium grained and less well cemented. This member is approximately 110 ft (33.5 m) thick. The Gibson Coal Member of the Crevice Canyon Formation overlies the Dalton Sandstone. The Gibson Coal Member consists mostly of inter-stratified sandstone, siltstone, and shale. The sandstone typically is fine grained and poorly cemented although some sandstone beds show planar cross stratification. The shale is largely dark colored and fissile. The combined thickness of the Gibson Coal Member and the overlying Hosta Tongue is approximately 130 ft (40 m).

The Hosta Tongue Member of the Point Lookout Sandstone overlies the Gibson Coal Member. The lower portion of the Hosta Tongue may be entirely sandstone whereas the upper few feet may be alternating thin to thick beds of sandstone and shale. The Hosta Tongue is basically fine to medium grained, well sorted, and moderately well cemented. It is mostly even bedded but some cross stratification is present. The Point Lookout Sandstone overlies the Hosta Tongue. This member is fine to medium grained, poorly sorted in its coarser parts, and fairly well cemented. Cleaner cross stratification is apparent in this unit that is approximately 132 ft (40 m) thick. These described formations constitute the units of interest in a hydrogeologic investigation of this mine. The nature of the lithology is indicative of hydraulic behavior if fracture permeability is not predominent.

The origin and localization of uranium in the Grants Mineral Belt is essentially undefined at this time (Melancon, 1963). A hypothesis for the formation of the San Juan Basin Mineral Belt in New Mexico was formulated by Granger (1968). Investigators have found that the individual ore bodies are blanketlike and elongate parallel to the sedimentary trends of the enclosing units. The primary ore occurs with the carbonaceous residue found in sandstones. This carbonaceous material has been mineralized to some degree with uranium, but no uranium of primary origin has been found isolated from carbonaceous material. The carbonaceous material coats sand grains and fills the openings between the grains and locally fills openings in or replaces highly altered detrital grains. A reducing environment is created around carbonaceous materials that is conducive to the precipitation of uranium. These materials are good adsorbents as well as being chemically conducive to combination with uranium-type organic compounds. Ore deposition was controlled by the presence of the carbonaceous material even if the two were not deposited simultaneously.

The primary uranium deposits in the Morrison Formation seem to occur in close association with areas where the structurally uplifted edge of the Morrison Sandstones were exposed at the paleo surface by early Cretaceous erosion. It is hypothesized that valuable carbonaceous material was carried down into the exposed Morrison Sandstone by precipitation. Mineralization therefore seems to be defined by the intersection of the paleo ground water table that was fluctuating in the uptilted Morrison aquifers. It is probable that the carbonaceous material concentrated the uranium and that it constitutes a primary control of ore deposition. Deposits of uranium occur in other formations in the Grants Mineral Belt; however, the Morrison Formation constitutes the host for the bulk of the uranium ore.

Mine History

The discovery of uranium in 1950 in the Grants Mineral Belt is generally attributed to a Navaho Indian prospector. However, uranium minerals were known to occur near Grants in the early 1920s (Melancon, 1963). The original discovery of uranium was in a limestone formation that underlies the Morrison Formation.

Initial work at the uranium mine of interest herein consisted of a hydrogeologic investigation for determining the parameters necessary for predicting shaft inflow. The techniques employed for analyzing formation characteristics such as hy-

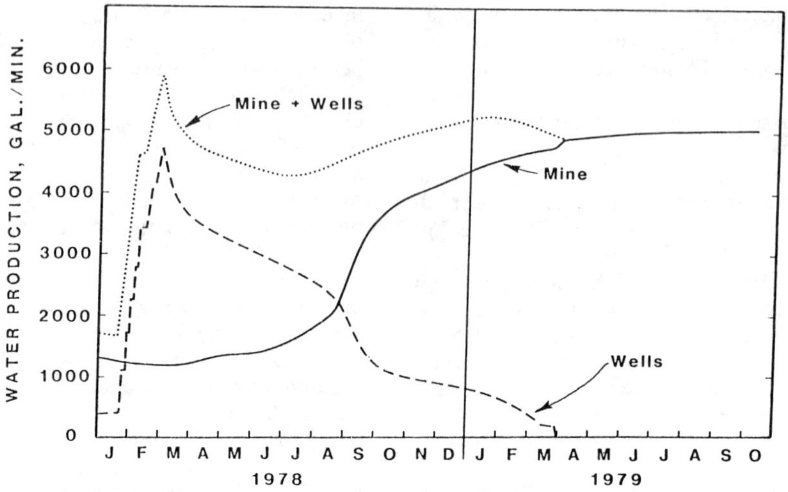

Fig. 69. Relationship of components of water produced at the uranium mine (supplied by the mining company).

draulic conductivity were developed by the petroleum industry. The sinking of the shaft at this mine was complicated by a large inflow of water. Only a 20-min lag time existed during one stage of development of the shaft between the time when pumps would shut down and the time when the shaft would flood. The shaft was completed along with about three miles of drift. The mine plan includes a modified room and pillar method for ore removal.

Pumpage from the mine at the time of this writing totaled about 5800 gpm (22 m^3/min), but only about 3000 gpm (11.4 m^3/min) is due to direct ground water inflow into the drifts. The balance of the pumpage is due to other water sources in the mine. Mine inflow has been essentially consistent with the increased area being mined. No sudden large inflows of water have occurred; the inflow has been consistently steady and high.

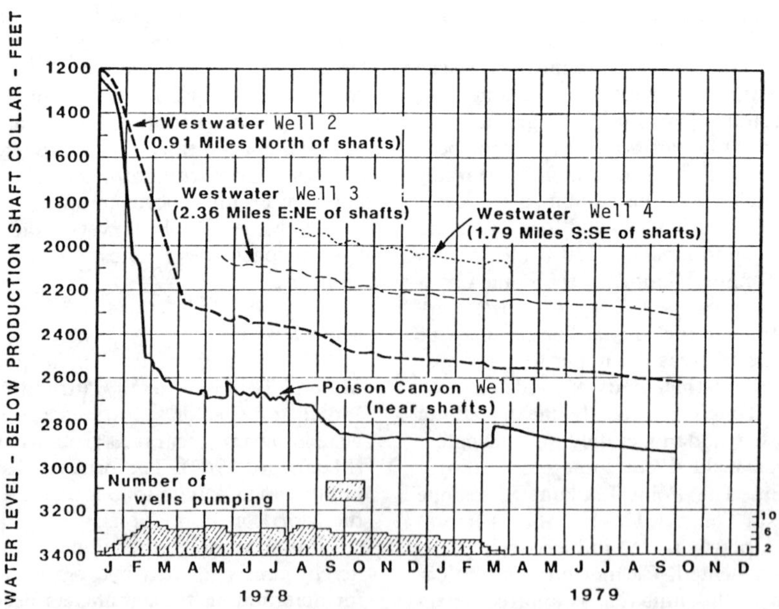

Fig. 70. Decline in water levels due to depressurization pumping (supplied by the mining company).

Data Collection

Surface Water System: No precipitation data are available for this site; however, precipitation for the area is low; it would be classed as semiarid to arid. No surface water data (stream flow or spring discharge data) are available.

Ground Water System: Water production data for the uranium mine were obtained in graphical form for the years 1978 and 1979. Fig. 69 shows the water produced by the mine and the depressurization wells both combined and separately.

Ground water level data also were obtained for periods of 1978 and most of 1979 as shown in Fig. 70. Ground water level data are available for wells in the Westwater Formation and one well in the Poison Canyon Formation. A location map for the Westwater wells is shown as Fig. 71. Fig. 72 is a graphical representation of the drawdown in the Westwater Formation as of October 1979 due to the depressurization around the shaft.

The initial hydrogeologic testing at this mine consisted of a single well drilled at the site of the proposed vertical mine shaft. This single well consisted of a 7-⅞-in. (20 cm) diam hole from which 4-in. (10 cm) cores were taken over the total depth. The overall core recovery was greater than 95%. The well was drilled using a water-in-oil emulsion drilling fluid due to the existence of poorly consolidated formations at the site. A 5-½-in. (14 cm) diam casing was installed in the well over its total depth of approximately 3200 ft (975 m). Selective cementing procedures were used to isolate those zones at the proposed shaft site that were tested. A cement bond log was run after the completion of all cementing operations to determine if the zones were isolated as desired. The casing was then perforated opposite those zones to be tested.

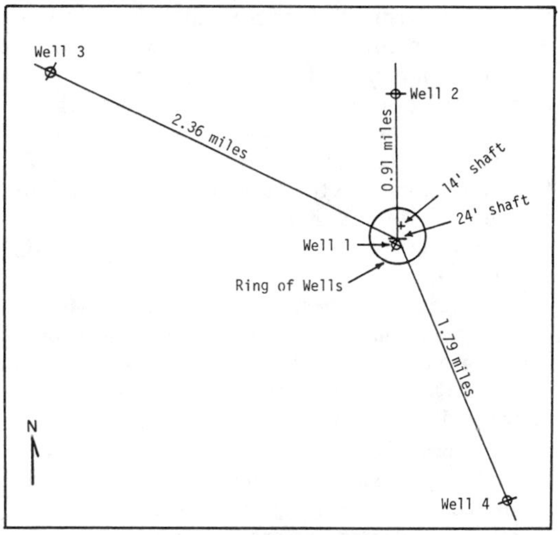

Fig. 71. Observation well locations (supplied by the mining company).

Four methods of analysis were selected by the company's mine hydrologist for evaluating the test data. The Horner method as described in Matthews and Russell (1967) was the method of analysis preferred by the mine hydrologist. The second method is referred to as the Maier method. This procedure is the creation of a service company that worked for the mining company at this site; the method is not documented. The third method of analysis is referred to as the McKinley method. The McKinley method utilizes a type curve procedure wherein

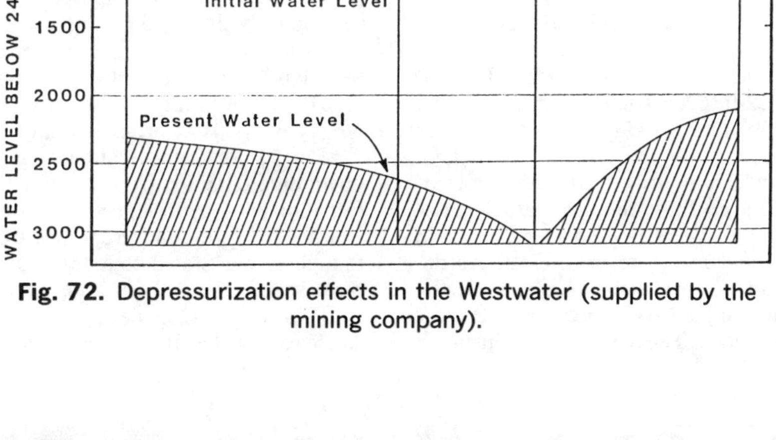

Fig. 72. Depressurization effects in the Westwater (supplied by the mining company).

data are superimposed on a family of type curves. The fourth method of analysis is referred to as a variable head method as used by the US Navy Bureau of Yards and Docks. According to the author of the company report, this method is described in Cedergren (1967, pp. 78-80). The foregoing tests form the basis for assigning relative transmissivities and storage factors to the formations and units of interest for shaft inflow prediction at this mine.

The results of these tests plus the other tests conducted at the site are summarized in Table 6. Values presented in Table 6 are in typical petroleum industry parameters such as transmissivity in millidarcies (feet per centipoise). A summary of all the data required for predicting shaft inflow is shown in Table 7. Porosity values shown in this table were estimated from core analysis and geophysical logs. Depth intervals and perforated intervals are noted in numbers of feet below the Kelly bushing. The Kelly bushing refers to the rotating base of a drill rig that was used for setting and pulling the test equipment.

Personnel involved with the early development of the uranium mine realized that an accurate estimate of formation characteristics at the shafts would be required in order to predict the long-term water influx during shaft sinking. Consequently the decision was made to conduct a second type of test at the site in order to define better the hydrogeologic properties of the three sand layers that would eventually be mined. These sand layers are the Poison Canyon Formation, the Upper Westwater Formation, and the Lower Westwater Formation. These formations had been tested previously using drill-stem type tests that, as noted by the investigators of the second report, do not test an area that extends to a sufficient distance from the test well.

The mine hydrologists decided to run a long-term constant rate pump test at the site. It was deemed desirable to obtain more accurate long-term pump test data for the prediction of long-term water inflow rates into the mine shaft. The investigators noted that hydrogeologic properties determined dynamically are more reliable than parameters determined from core and geophysical log data. A long-term test also reduces any potential dominating effect created by a fracture system in the sand formations that would eventually be mined. The originally proposed 22-ft (6.7 m) diam shaft was revised to consider a 24-ft (7.3 m) diam shaft for further evaluation.

A new test well was drilled at a site approximately 51 ft (15.5 m) from the well tested previously. This new well was designed to gather data from the three sand members of interest. As described earlier, these three sand members are the Poison Canyon, Upper Westwater, and Lower Westwater Formations. The new well was installed with a submersible pump and two pressure sensing devices. The well used for drill-stem tests was modified so as to monitor only pressure changes due to pumping in the three sand layers of interest. This well was also equipped with a pressure sensing device.

The submersible pump used for this pump test was driven by an on-site generator. Discharge from the pump was controlled by an automatic control valve and measured with an electromagnetic flowmeter. Difficulties were encountered in both starting the pump for the pump test and maintaining a constant rate of discharge.

The first pump test was conducted after observing static pressures in both the test and observation wells for several hours before the test was begun. The problems inherent in conducting such a test in deep holes can be numerous. Those encountered in this test are discussed in detail in order to illustrate some of them. The pressure sensor installed below the pump malfunctioned before the test began so that no pressure record was available from this pressure sensing device. The test was conducted with one pressure sensor in the test well although an attempt was made to collect pressure data in the observation well. This attempt proved unsuccessful due to tubing leaks, malfunction of the surface recorder, and malfunction of the slim hole sensor installed in the well. An incomplete and highly questionable pressure record was obtained. The mine hydrologist did not use data from the observation well for any calculations or analysis of hydrogeologic properties. The pumping rate that was to be regulated and held constant by a flow controller was also fraught with problems. The controller did not function properly during the first two pump test cycles. The rate control equipment did function properly, for a third pumping cycle. The variation in pumping rate during the first two pumping cycles rendered the drawdown data unusable; however, the pressure buildup or recovery data were usable according to the investigator. The third pumping cycle provided good drawdown curve data. The observed time drawdown data are presented in Fig. 73. The pump test discharge rate is shown in Fig. 74. Pump discharge was recorded continuously during these tests.

Data from these pump tests were analyzed using techniques described in Matthews and Russell (1967). The techniques used are petroleum technology methods of pressure analysis.

Fig. 75 illustrates the long-term pressure drawdown data obtained from the third cycle pumping test. The mine hydrologist did not present any calculations of hydraulic properties from this plot although it was stated that this plot indicates that the reservoir is nearly infinite in size.

Pressure buildup data or recovery data were analyzed for the first, second, and third buildups

Table 6. Calculated Transmissivities, Uranium Mine (Supplied by the Mining Company)

Interval				Calculated Transmissibility Values,[§] md-ft/cp						
No.	Flow period	Perforated interval ft, KB	Formation name	Horner	Maier	McKinley	Variable head	Injection	Productivity index	Most likely
1b	3	3161–3231	Lower Westwater	No data	30,000	26,000	30,000	No data	20,000	30,000
2	3	3093–3135	Upper Westwater	No data	26,000	27,000	20,000	No data	20,000	25,000
	4			No data	24,000	27,000	18,000	No data	20,000	
3	1	3024–3073	Upper Poison Canyon	No data	85,000	100,000	75,000	No data	70,000	125,000
	3			No data	130,000	120,000	70,000	No data	80,000	
	4			No data	No data	130,000	70,000	No data	No data	
4	1	2885–2952	Dakota Brushy Basin	2400	1,900	2,500	1,800		1,600	2,500 to 5,700
	3			2900	2,500	2,400	2,200	5,700*	2,200	
	4			2400	No data	No data	2,700		3,000	
5	1	2765–2859	Lower Tres Hermanos	No data	1,600	2,500	1,700	No data	1,800	2,500
	3			No data	No data	No data	2,500	No data	2,800	
	4			No data	2,300	No data	2,400	No data	2,700	
	5			No data	2,300	No data	2,700	No data	2,200	
6	1	2690–2725	Middle Tres Hermanos	4	No data	6	10	No data	13	10
7	1	2565–2623	Upper Tres Hermanos	25	No data	No data	No data	No data	40	25
9	1	1980–2005	Lower Gallup	300	No data	No data	No data	No data	300	300–1500
10	1	1752–1840	Gallup	1600	2,300	1,700	2,300	No data	2,500	2,500
	3			2400	No data	No data	3,100	No data	3,100	
	4			2500	2,500	2,200	2,100	No data	3,300	
	5			No data	2,000	2,300	2,300	No data	2,400	
11	1	1648–1712	Stray-Dilco	No data	290	370	330	No data	270	300
	3			No data	No data	280	340	No data	400	
	4			No data	No data	290	330	No data	370	
12	1	1303–1348	Mullato Tongue	No data	510	370	700	No data	510	600
	3			No data	No data	370	720	No data	540	
	4			No data	600	370	710	No data	600	
	5			No data	600	550	710	No data	600	
13	1	1147–1237 1101–1111	Dalton	1600	2,000	2,100	2,300	No data	2,500	2,000
14	1	920–1012	Gibson	No data	6,900	8,500	6,600	No data	7,100	7,500
	3	1025–1045	Lower Hostra	No data	No data	No data	6,800	No data	6,500	
	5			No data	7,100	No data	7,200	No data	6,900	
	6			No data	7,500	8,500	7,200	No data	6,700	
15	1	788–894	Point Lookout	No data	No data	26,000	No data	54,000[†]	30,000	30,000
	2			No data	34,000	32,000	24,000		25,000	
	4			No data	29,000	22,000	22,000	34,000[‡]	40,000	
	5			No data	30,000	26,000	33,000		40,000	

* From pressure fall-off after injection.
[†] From pressure fall-off after injection; not reliable, fall-off time too short.
[‡] From two-rate injection test.
[§] No data denotes either, no test run, insufficient data taken, or data was uninterpretable.

Table 7. Measured and Calculated Properties of Hydrostratigraphic Units at Proposed Mine Shaft (Supplied by the Mining Company)

Formation name	Depth interval, ft, KB *1	Perforated interval, ft, KB *1	Gross thickness, ft	Net thickness, ft *2	Average porosity, fraction *3	Static water pressure, psig (at midpoint of interval) *4	Piezo-metric surface ft, KB *1	Trans-missi-bility, md-ft/cp *5	Average perme-ability, md *6
Lower West-water	3152–3240	3161–3231	88	84	0.13 (0.22)	942	1020	30,000	200
Upper West-water	3088–3140	3093–3135	52	50	0.12 (0.23)	902	1030	25,000	290
Poison Canyon	3020–3081	3024–3073	61	55	0.16 (0.25)	875	1030	125,000	1,300
Dakota Brushy Basin	2878–2961	2885–2952	83	75	0.15est. (0.22)	817	1030	2,500 to 5,700	25 to 55
Lower Tres Hermanos	2753–2871	2765–2859	118	118	0.15 (0.22)	790	980	2,500	13
Middle Tres Hermanos	2685–2730	2690–2725	45	40	0.16 (0.23)	766	940	10	0.2
Upper Tres Hermanos	2558–2630	2565–2623	72	58	0.12 (0.19)	751	860	25	0.3
Lower Gallup	1980–2010	1980–2005	30	20	0.17 (0.24)	662	470	300 to 1,500	10 to 50
Gallup	1742–1850	1752–1840	108	50	0.22 (0.29)	581	460	2,500	35
Stray-Dilco	1640–1720	1648–1712	80	36	0.21 (0.31)	523	470	300	6
Mulatto Tongue	1297–1354	1303–1348	57	44	0.12 (0.28)	420	460	600	10
Dalton	1138–1248	1147–1237 1101–1111	110	100	0.20 (0.35)	346	400	2,000	20
Gibson-Lower Hostra	920–1050	920–1012 1025–1045	130	130	0.22 (0.35)	306	280	7,500	40
Point Lookout	775–907	788–894	132	130	0.20 (0.35)	243	280	30,000	170

*1 is KB–Kelly bushing.
*2 is from core description and well logs.
*3 is geophysical log data in parentheses.
*4 is from flow test data.
*5 is millidarcy-feet/centipoise.
*6 is millidarcy.

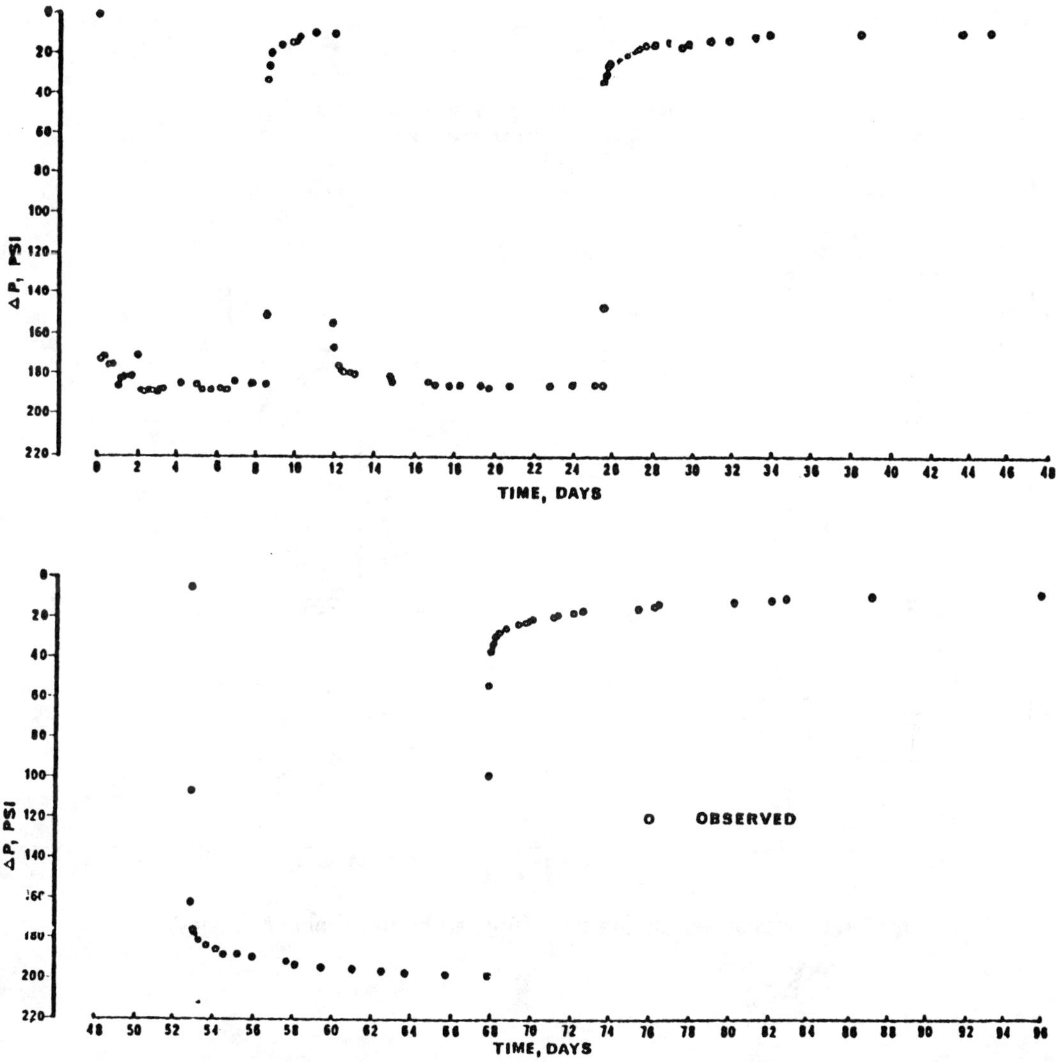

Fig. 73. Pressure decline due to pump test (supplied by the mining company).

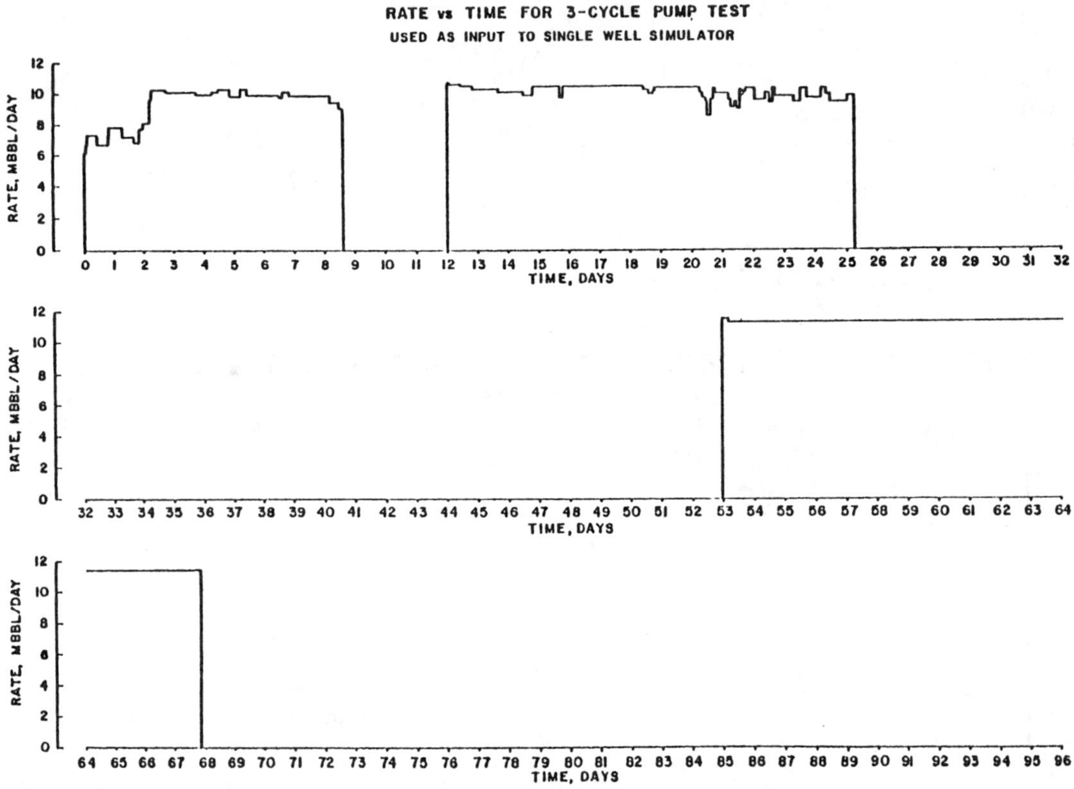

Fig. 74. Pumping rate during test (supplied by the mining company).

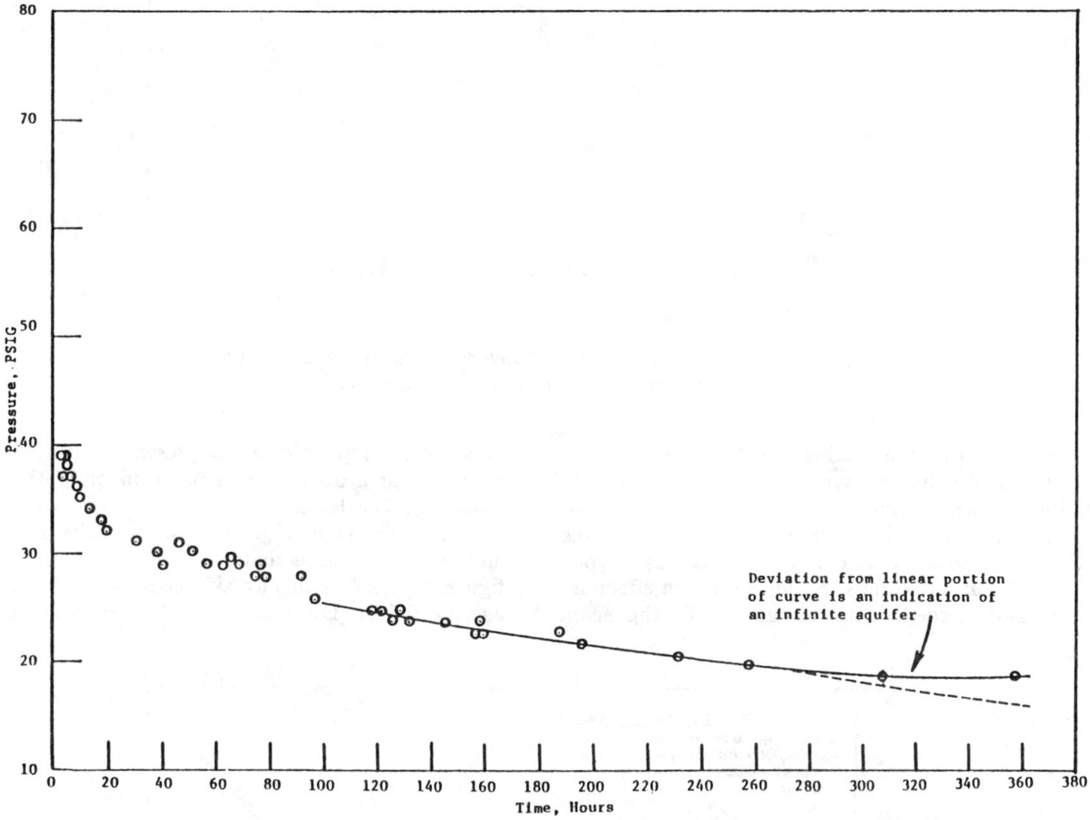

Fig. 75. Pressure drawdown analysis for semi-steady state conditions (supplied by the mining company).

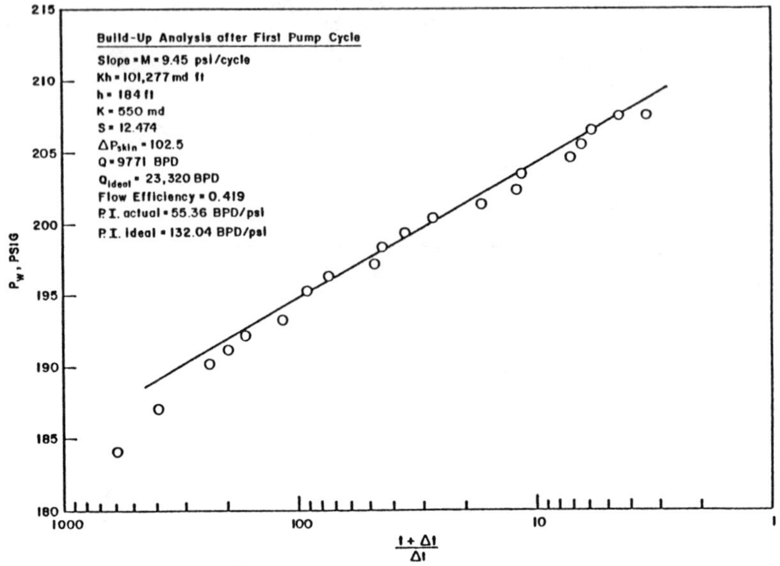

Fig. 76. Horner plot for first recovery period (pressure buildup) (supplied by the mining company).

from the pump tests. Fig. 76 illustrates the data plot for the first pressure buildup analysis; Fig. 77 illustrates the second pressure buildup. The Horner method of analysis was used to calculate transmissivity as *Kh* and hydraulic conductivity as *K* where *h* is the unit thickness. *S* stands for skin effect and not storage coefficient. The data plot for the second pressure buildup indicates the presence of a barrier boundary at a distance of approximately 7900 ft (2408 m). The Horner plot for the third pressure buildup is shown as Fig. 78. Transmissivity as *Kh* and hydraulic conductivity as *K* are shown in this figure. Again, *S* stands for skin effect and not storage coefficient. Distance to the barrier boundary

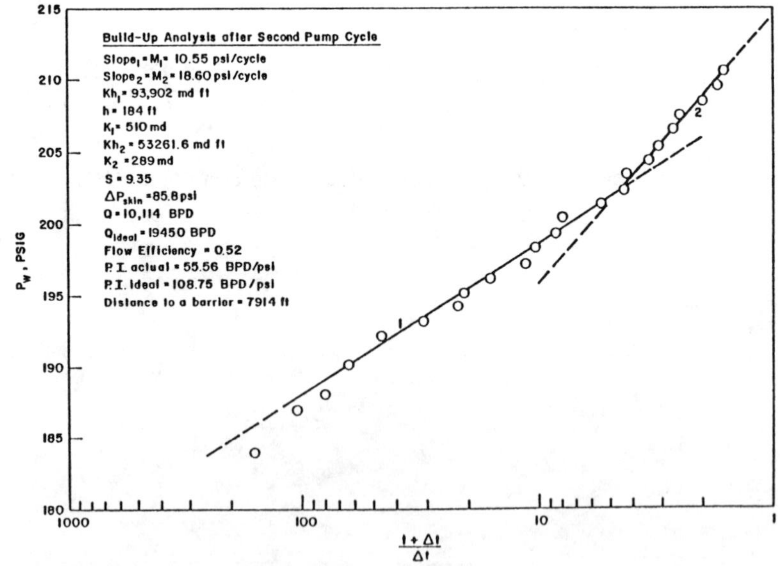

Fig. 77. Horner plot for second recovery period (pressure buildup) (supplied by the mining company).

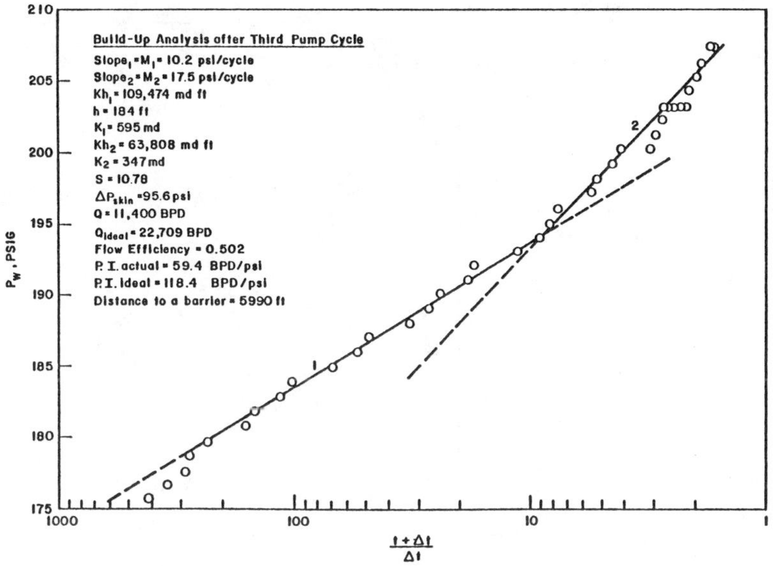

Fig. 78. Horner plot for third recovery period (pressure buildup) (supplied by the mining company).

noted previously was calculated as 5990 ft (1826 m) on this plot.

Data from this test and the previous drill-stem test were incorporated into a single well model developed as a computer program by the mining company. This model is a numerical, two-dimensional, multiphase reservoir model formulated for a radial-vertical coordinate. The model used in this instance was single phase rather than multiphase. The model considers the effects of fluid and formation compressibility, gravity, and convergent flow into the well bore. It also allows for variable formation and fluid properties. This model was used to simulate the third cycle drawdown test since this was the only test that had a constant rate discharge. This attempt to simulate the third cycle pump test indicated that the permeabilities were too high as placed in the input to the model. Transmissivity subsequently was redetermined from data derived in the drawdown portion of the third cycle of the pump test. Permeabilities or hydraulic conductivities were again determined with iterative calculations. Skin effect (that is a reflection of well losses at the pumping well) was estimated from earlier calculations. A good match of historical data was then obtained with this simulation as shown in Fig. 79. This computer verification by the mining company indicates that the Poison Canyon Formation has a permeability of 1029 millidarcies, the Upper Westwater Formation has a permeability of 380 millidarcies, and the Lower Westwater Formation has a permeability of 193 millidarcies. Porosity for the reservoir was set at 0.19. The system compressibility was set at 6.75 x 10^{-6} psi (4.65 N/m^2). The computer simulation of the third cycle pump test data vs. observed data is shown in Fig. 79. The mining company computer simulation for the full three cycles of the pump test is shown in Fig. 80. The increasing similarity of their computer simulation with observed data was interpreted by the mine hydrologist as indicating that the formations outside the perforated portion of the test well were being cleaned up gradually due to the pumping. In other words, the original formation characteristics that had been altered during the drilling operation were being returned nearly to their original values.

Inflow Prediction by the Mining Company

Data from the original drill stem test at this site were analyzed for the various hydrogeologic properties needed for predicting shaft inflow. The shaft inflow was hand calculated by two techniques, neither of which is documented. The first technique as illustrated in Table 8 indicates the predicted inflow from each of the various hydrostratigraphic units for various time periods. The second method of analysis for predicting shaft inflow is based on the Theis equation described previously. Inflows were predicted for each of the hydrostratigraphic units. The highest inflow rate from a unit was estimated to be 4900 gpm (18.5 m^3/min) while the lowest was approximately 1 gpm (0.004 m^3/min).

Inflow prediction techniques for the second phase of testing were accomplished in two phases.

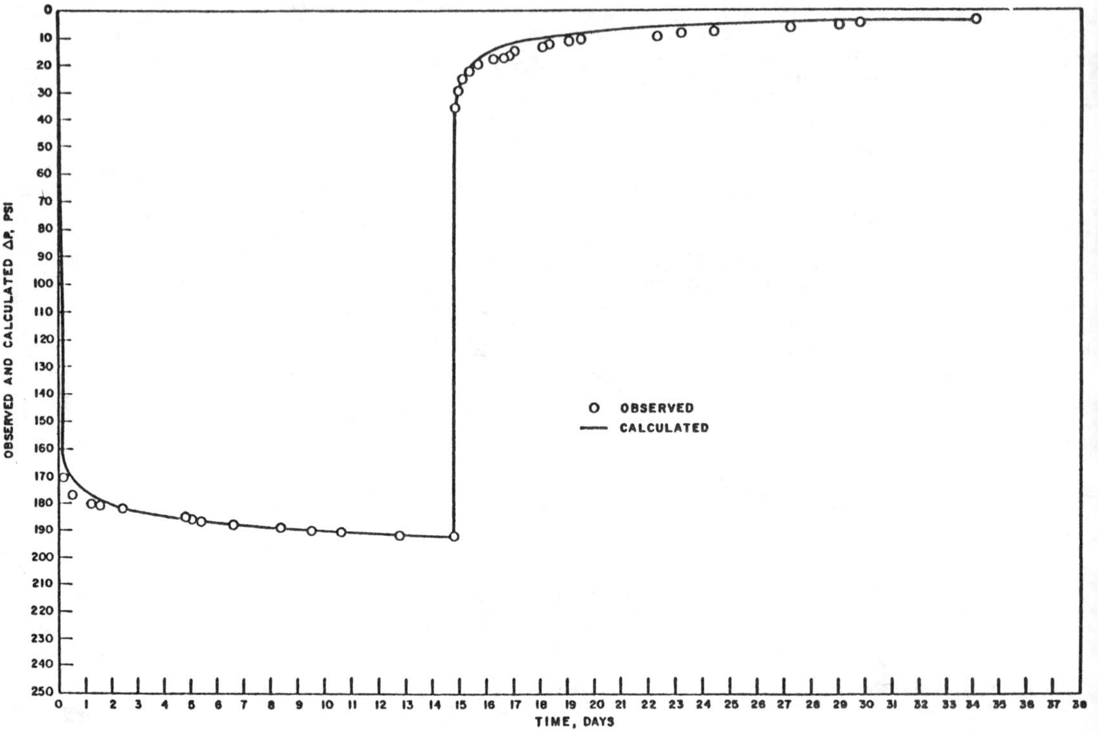

Fig. 79. History match of third pumping cycle Δp to calculated Δp (supplied by the mining company).

The first phase of the mining company predictive techniques utilized the two-dimensional computer model noted previously. Inflow was predicted by the mining company for vent shaft inflow based on a single well case. A computer simulation was run with the single well model utilizing previously noted hydrogeologic parameters and assuming a constant head boundary at 18,000 ft (5486 m) from the pumping well. The computer simulation of pressure declines is shown in Fig. 81. It should be noted that the simulation from the time period 0.55 days does not represent a maximum drawdown due to the problems encountered by having the simulator handle the huge change of mass created by the maximum drawdown. The production rate required to simulate the required pressure drawdown at this single well model was 15,819 m^3/d (2900 gpm) with a stabilized gage pressure in the shaft of 54 psi [125 ft (372 kN/m^2)].

A second simulation using this two-dimensional model was run by the mining company using the assumption that the pressure was drawn down to zero at the top of the sand layer (Fig. 82). Again, a constant pressure boundary was set at a distance of 18,000 ft (5486 m). These simulated pressure drawdowns for the single well model are shown in Fig. 82. The stabilized discharge rate from the single well model was 16,900 m^3/d (3100 gpm) with a stabilized pressure of 0 psi gage at the sand top. The two-dimensional model predictions were verified by the mining company using an unspecified gas engineering hand calculation technique from Katz (1959).

A three-dimensional model was selected by the mining company hydrologist to model the two aforementioned dewatering points represented by the two proposed shafts separated by approximately 350 ft (107 m). The model selected was designed to simulate two-phase flow systems and can be used to study numerous systems of compressible reservoir fluid flow. This model considers variations in porosity, horizontal and vertical permeabilities, and also arbitrary cross-sectional geometry. Fluid properties in the model are a function of pressure and the simulator incorporates the effect of relative permeability, gravity, and viscous forces. The model was operated for single-phase flow for this simulation. The same hydrostratigraphic unit properties incorporated in the two-dimensional model were also used in this three-dimensional model. The shape of the system modeled three dimensionally is a square with an area equal to the area of the

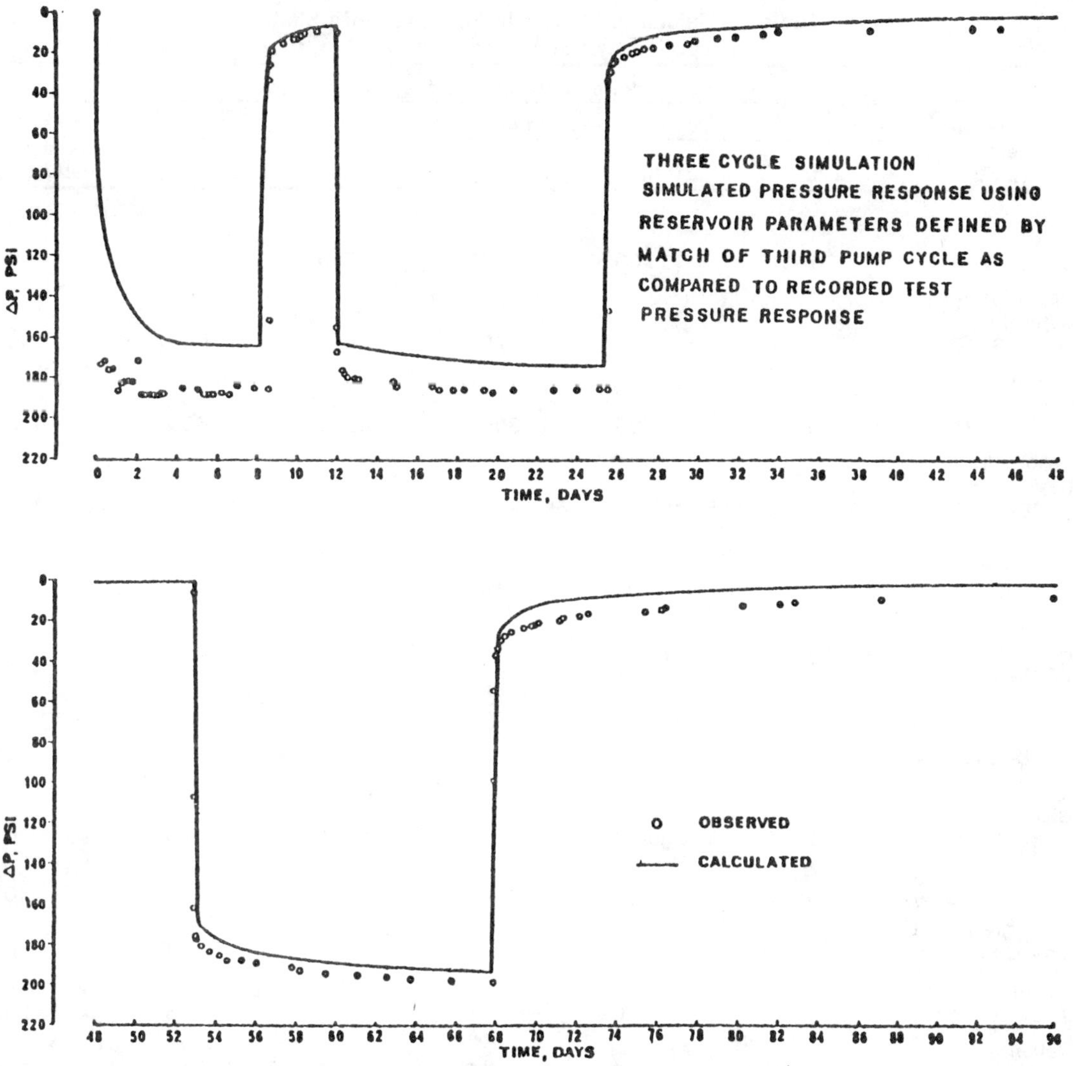

Fig. 80. History match of all pumping cycles Δp to calculated Δp (supplied by the mining company).

18,000-ft (5486-m) radius circle used for the two-dimensional model. The constant pressure boundary concept utilized in the two-dimensional model was not used in this case. It was decided to simulate the in situ aquifer conditions by setting up a row of injection wells at a specific location on the edge of the grid system. The amount of water produced from the shaft locations for each time step was injected into this row of injection wells.

The conditions simulated by the mining company are as follows: The vent shaft was considered to be excavated to a depth of approximately 2875 ft (876 m) where a drilling station was to be established. A ring of dewatering holes would be drilled outside the grout curtain around the perimeter of the shaft and a temporary pump station would be installed at this 2875-ft (876-m) level. Dewatering from this level would begin and continue as the shaft was sunk through the underlying Westwater Formation. Dewatering operations would be expanded to include another dewatering point at the main shaft after six months of dewatering at the vent shaft. It was assumed that several aspects of the shaft sinking operations would be completed during this six-month period. The main shaft sinking would be in progress and the vent shaft would be completed through the Westwater Formation. A drainage drift would be driven from this location at the vent shaft below the Westwater Formation to the bottom hole location of the main

Table 8. Calculated Water Influx Rates into Shaft (Supplied by the Mining Company)

Formation depth, ft, KB	Water influx into 22-ft diam. shaft, gpm					
	Maier method of calculation					Theis-Deju method
	1 hr	10 hr	100 hr	1000 hr	10,000 hr	
Lower Westwater 3152–3240	1400	1100	900	750	650	1400
Upper Westwater 3083–3140	1100	900	750	650	550	1100
Poison Canyon 3020–3081	5600	2700	3500	2900	2600	4900
Dakota-Brushy Basin 2878–2961	100 to 200	80 to 180	65 to 150	55 to 130	50 to 110	120 to 250
Lower Tres Hermanos 2753–2871	70	50	40	35	30	10
Middle Tres Hermanos 2685–2730	1	1	1	1	1	1
Upper Tres Hermanos 2558–2630	2	2	1	1	1	2
Lower Gallup 1980–2010	8 to 40	6 to 30	5 to 25	4 to 20	4 to 16	10 to 50
Gallup 1742–1850	70	50	40	35	30	90
Stray-Dilco 1640–1720	8	6	5	4	4	10
Mulatto Tongue 1297–1354	13	10	8	7	6	20
Dalton 1138–1248	35	30	20	15	15	50
Gibson-Lower Hostra 920–1050	115	90	70	60	50	130
Point Lookout 775–907	350	270	220	180	160	410

shaft 350 ft (107 m) away. Drainage holes would then be drilled up into the Westwater Formation at the bottom hole location of the main shaft with the ring of holes located on a 20-ft (6-m) diam circle.

The simulation scheme was changed to add the second dewatering point at the main shaft location with complete drawdown occurring at both shaft locations. Complete drawdown was defined as 0 psi gage at the sand top depth of 3000 ft. The simulation using the three-dimensional model was continued for an additional eight months to complete a two-year run.

Results of the three-dimensional mining company simulation are illustrated by Figs. 83 and 84. Fig. 83 illustrates the pressure profile through both shaft locations. This figure illustrates the initial drawdown created by the vent shaft dewatering at early times as well as the additional drawdown created by dewatering at the main shaft at later times. Fig. 84 illustrates the pressure profile through the vent shaft on a line perpendicular to

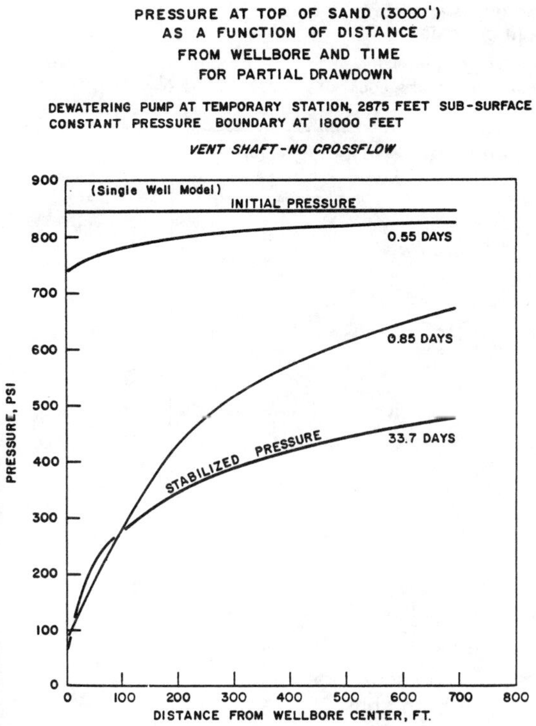

Fig. 81. Simulated decrease in pressure for partial drawdown due to depressurization pumping (supplied by the mining company).

a line joining both the shaft locations. Again this figure illustrates the declining pressures with time at this location. The steady state production rate from the vent shaft for the first six months was 14,834 m^3/d (2721 gpm) with partial drawdown. The steady state pressure at the sand top in the shaft was 54 psi [125 ft (372 kN/m^2)]. The steady state combined production rate from both shafts after six months was 20,589 m^3/d (2777 gpm) with the pressure at each shaft location being 0 psi gage that corresponds to the complete drawdown imposed upon the system.

A significant conclusion reached by the mining company during the course of the pump test and in the analysis of the pump test data was that there was no dominant fracture system affecting ground water flow at the uranium mine. Consequently, the aquifer behaves as a common sandstone reservoir in petroleum terminology or, in other words, the system behaves as a porous medium wherein interconnected interstitial pore spaces control hydraulic conductivity. Our field site investigation verified this conclusion. The mining company investigators who conducted the second set of hydrogeologic tests at this mine noted that significant unknown parameters still exist with respect to making detailed predictions on dewatering requirements. These unknown parameters are the size, shape, and rate of recharge to the aquifer. These investigators noted that their predicted rates of inflow at least established the order of magnitude which could be expected.

The model was used also by the mining company to simulate the anticipated total water that would have to be pumped from the mine. This simulation is illustrated in Fig. 85. The model also has been used to simulate the projected rate of inflow to the drift vs. the drift length based on the drifting rate. This relationship is illustrated in Fig. 86. A simulation also was made to determine the water inflow rate per foot of drift as a function of the location of the drift from the point of origin. This relationship is illustrated in Fig. 87 that shows the rate of water inflow to the drift vs. the drifting rate as distance from the drift origin.

As can be seen by the previous summary of work done at the mine, a significant amount of testing and analysis has been conducted on the hydrogeologic parameters. Shaft inflow prediction methods included both hand calculated procedures and computer simulations. The computer simulations were both two- and three-dimensional. The computer simulation has been used also to estimate

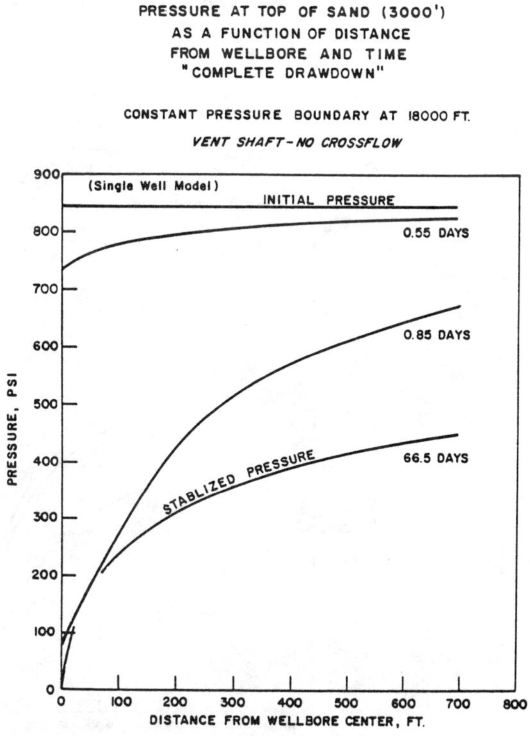

Fig. 82. Simulated decrease in pressure for complete drawdown due to depressurization pumping (supplied by the mining company).

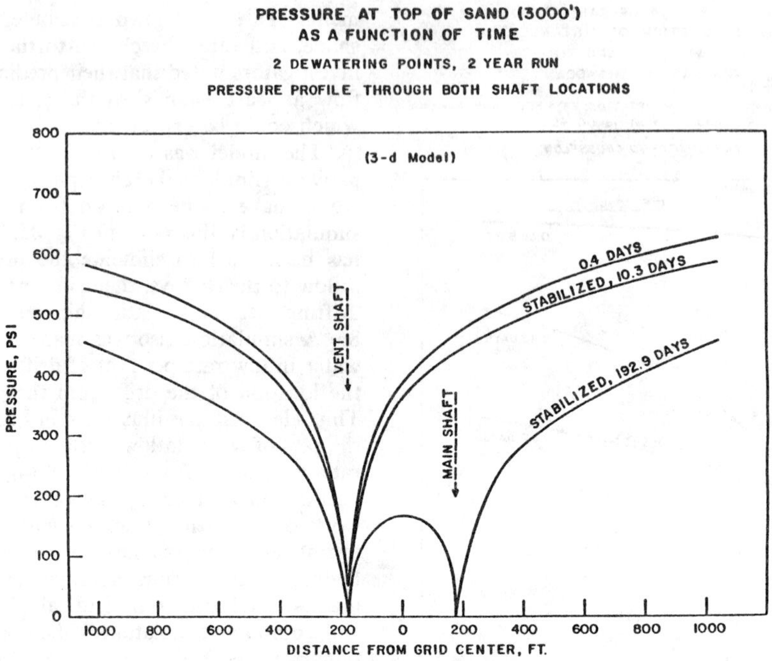

Fig. 83. Simulated decrease in pressure for two-year simulation through both shaft locations (supplied by the mining company).

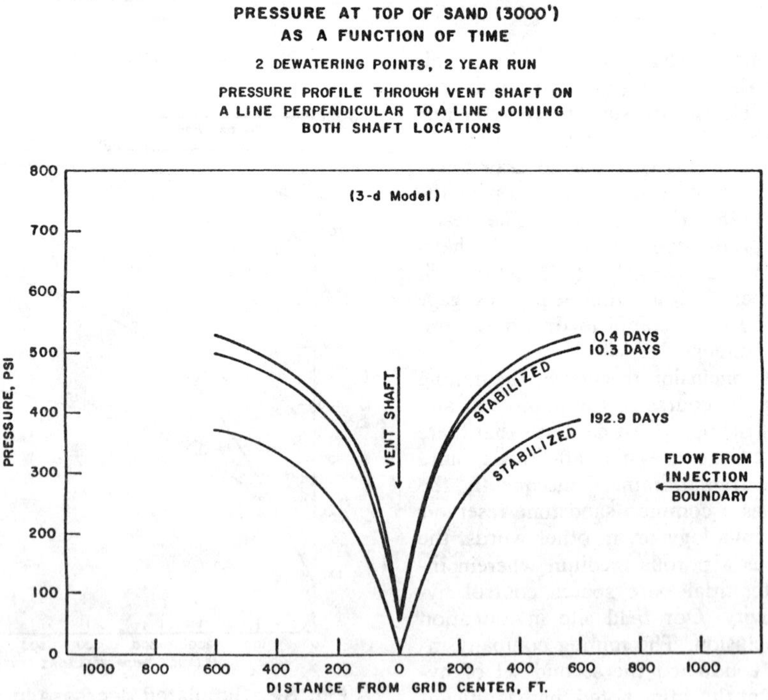

Fig. 84. Simulated decrease in pressure for two-year simulation (supplied by the mining company).

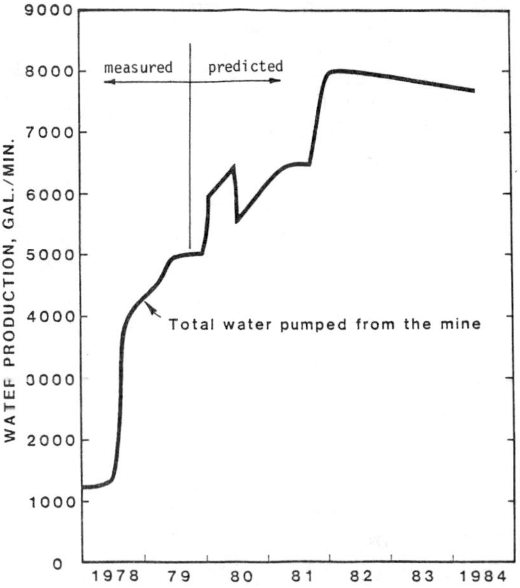

Fig. 85. Past and predicted future water production (supplied by the mining company).

inflow rates into the drifts as drifting progresses away from the point of origin at the shaft.

Several miles of drift existed at the time of our work. This mileage will increase significantly as the ore deposits away from the current site are extracted.

Simulation of Uranium Mine Water Inflow by the Authors

Simulation of Shaft Inflow: We elected to use the finite element model UNSAT2 to simulate water inflow to the uranium mine so as to duplicate various parts of the hydrologic study from which head and discharge data were available in the data supplied by the mining company. The first step in our simulation was to develop a finite element mesh to describe our conceptual model of the field situation. This procedure by necessity is an approximation to some extent because it is impossible to describe all the details of a geologic framework in a form that can be transferred to a mesh. For this mine it was decided to use the Poison Canyon and Upper and Lower Westwater Formations in the simulation. These were the formations used by the mining company for simulation. The combined thickness of the three formations is about 220 ft (67 m). Our finite element mesh shown in Fig. 88 was applied. This mesh has a depth of 220 ft (67 m) and a length of 18,000 ft (5486 m). This mesh could be used for an axisymmetric simulation where the left side is the axis of symmetry and the axis of the shaft. A two-dimensional analysis of the flow into a drift can utilize the same mesh; the drift would appear as two node points at the left side of the mesh.

The potential throughout the entire flow field was established at 2210 ft (674 m) above the bottom of the formation as the initial condition. The boundary condition at the outer extent of the mesh is a constant potential. This condition is the same as that used by the mining company. The values of hydraulic conductivity, porosity, and storage coefficient were taken from the computer-simulated historical match. The hydraulic properties (capillary pressure-saturation-relative conductivity relationships, Appendix E) of the sandstone materials were taken from White (1968), for sandstones that had

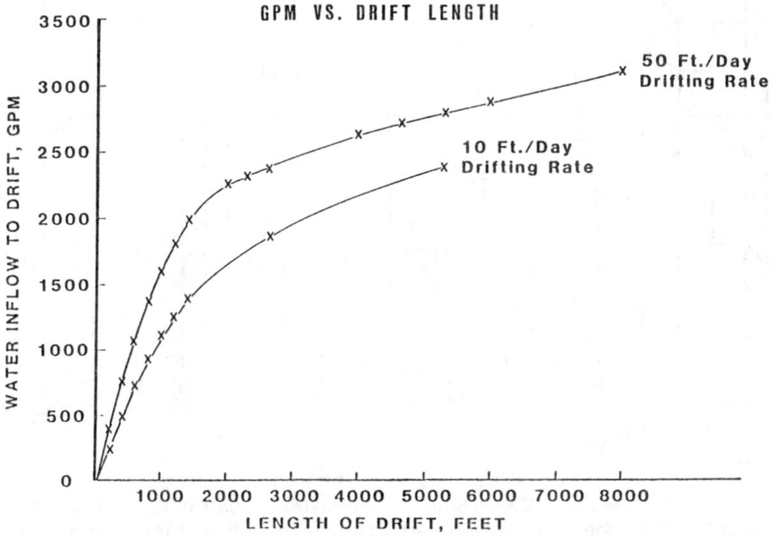

Fig. 86. Simulated rate of inflow into drift vs. drift length (supplied by the mining company).

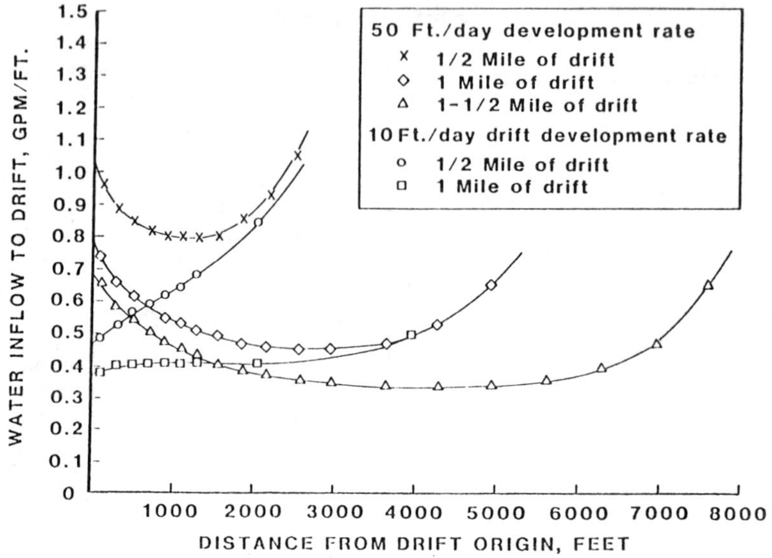

Fig. 87. Simulated drift inflow per foot of drift as a function of location (supplied by the mining company).

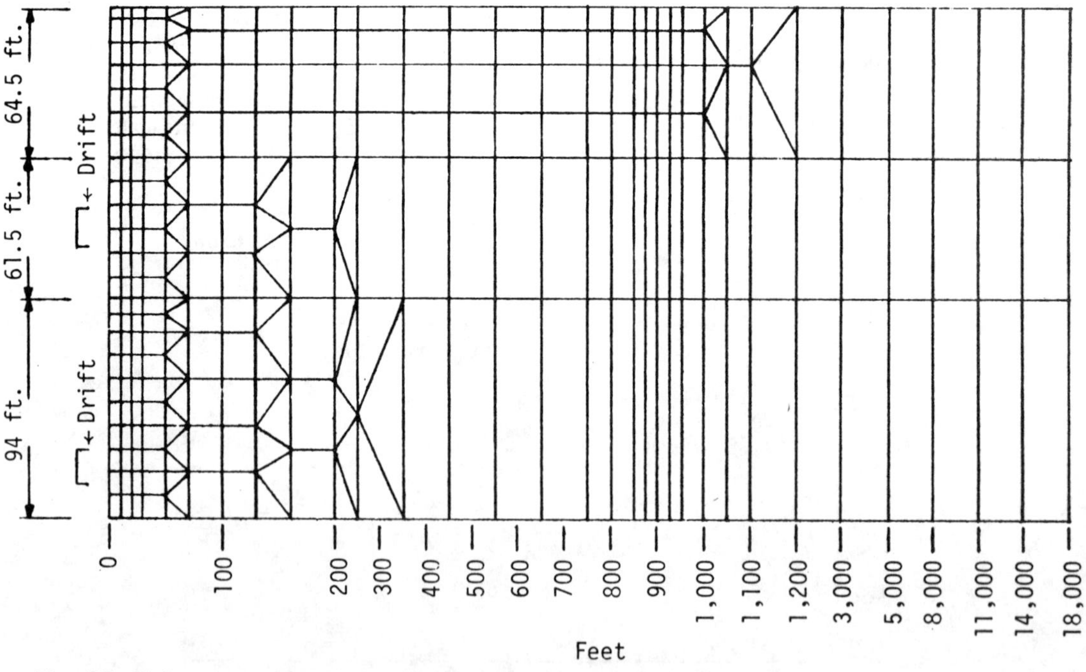

Fig. 88. Finite element mesh for the uranium mine. Note: Mesh is not drawn to scale. Coarse elements on right side of mesh are elongated laterally to reduce total number of nodes while expanding the mesh to 18,000 ft (5486.4 m) horizontally.

similar values of porosity and hydraulic conductivity. This procedure will give values at least within ±50% of the true value. However, in this problem the flow is nearly all saturated flow, in which case the unsaturated hydraulic properties are not used in the solution. Table 9 presents all values for the materials which we used in the simulation.

As discussed previously, the mining company did a limited amount of simulation on this problem in the history of the third pumping cycle shown in Fig. 78. The agreement between their observed and calculated values was very good. This was the first problem for which simulation was attempted. Our simulation results, however, were unsatisfactory. A great deal of time was spent on this particular problem with no satisfactory results. When the personnel with the mining company that had done the work were contacted, it was determined that they apparently had found errors in their modeling effort; they then changed the values of input hydraulic conductivity significantly in order to correct the results. They also had learned from some of their pump test analyses that there appeared to be a boundary affecting the flow into the wells. The boundary did not appear to be impermeable but simply boundaries of regions in which the conductivity was significantly less than that used for the Poison Canyon and Westwater Formations. The hydraulic conductivity of the formations also decreased as the distance from the shaft increased.

The mining company also used their axisymmetric model in an additional shaft inflow simulation to predict the drawdown in a single well at a distance of 13.6 ft (4.1 m). This simulation was run for a period of 0.55 days, 0.85 days and 33.7 days as originally shown in Fig. 81 and repeated in Fig. 89. The discharge which they used was 2900 gpm (11 m^3/min). Some factors about their data are confusing: (1) The drawdown in the first period was about 250 ft (76 m); then during the next 0.3 days the drawdown increased another 1500 ft (457 m); then during the next 33 days the drawdown increased another 60 ft (18 m). (2) The drawdown curves for 0.85 days and 33.7 days cross; this does not seem reasonable. We ran a simulation attempting to duplicate the conditions that the mine hydrologist used and produced considerably different curves. Our data appear to be defensible because our curves do not cross and they show that initially there is a large drawdown; the rate of change of this drawdown decreases with time. Subsequently, the company's mine hydrologist informed us that they had concluded that their simulation was incorrect.

Our next attempt at simulation was directed at the data shown in Figs. 69 and 70. These data give the pumping rate from the wells and the mine shaft and the resulting drawdowns during the time immediately before the shaft entered the Poison Canyon Formation and during the shaft construction.

Table 9. Values of Material Properties Used for Uranium Mine Simulation

Saturation, S	Moisture content, θ	Capillary pressure, ft of water, ψ	Relative conductivity, K_r
	Poison Canyon Formation		
0.059	0.013	17.2	0.13E-09
0.065	0.015	10.9	0.209E-07
0.110	0.025	6.22	0.107E-04
0.162	0.036	4.66	0.267E-03
0.383	0.086	3.08	0.027
0.903	0.202	2.33	0.599
0.916	0.205	2.17	0.800
0.929	0.208	2.01	0.900
1.0	0.224	0	1.0
	Upper and Lower Westwater Members		
0.426	0.079	44.9	0.31E-05
0.523	0.097	23.2	0.56E-03
0.571	0.106	17.7	0.0048
0.642	0.119	14.1	0.029
0.807	0.149	10.9	0.217
0.882	0.163	9.99	0.432
0.987	0.176	8.59	0.750
0.993	0.182	7.18	0.900
1.0	0.185	0	1.0

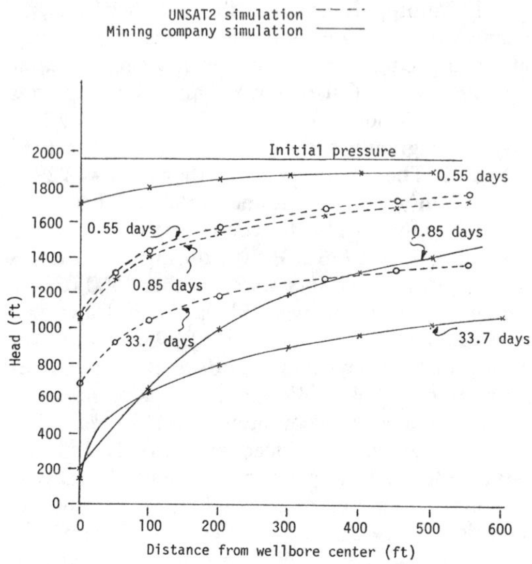

Fig. 89. Simulated drawdown in a well.

The mining company apparently did not attempt to simulate this process. The depressurization wells are located in a ring at a radius of about 900 ft (274 m) from the location of the shafts (Fig. 88). In our axisymmetric simulation the shaft location is considered an axis of symmetry. The depressurization wells, therefore, appear as a point source in our finite element mesh. Consequently we approximated the ten wells in a ring by a circular line sink at a radius of 900 ft (274 m). It was expected that this would cause some discrepancy between the data and the simulation when only a few wells were pumping. To simulate the boundary that the mining company reported, we inserted a material with a lower hydraulic conductivity in all elements at a radius larger than 3000 ft (914 m).

To begin simulation of the data in Figs. 69 and 70, it was necessary to pump the center shaft at an unknown rate to develop a piezometric height of 1966 ft (599 m) at node 194. Node 194 best represents observation well 1. Considerable trial-and-error was required to develop this drawdown since the pumping rate was unknown. Ultimately we used a pumping rate of 1100 gpm (4.2 m^3/min) for a time of 10.3 days. At this time the center shaft was shut off and all the water pumped was derived from the ring of depressurization wells. During approximately the first 20 days a discrepancy appeared in the discharge measurements. Fig. 69 shows the discharge from the three wells as 400 gpm (1.5 m^3/min) whereas the information on the number of pumps operating (from the bottom of Fig. 70) indicates that up to three pumps were operating that should produce a discharge on the order of 1100 gpm (4.2 m^3/min). After numerous trials, we finally decided to use 800 gpm (3 m^3/min) for a period of 20 days. The actual pumping rates as given in Fig. 70 were then used for about 50 days during which the agreement between the simulation and the observations in the field became reasonable. At this time the potentiometric surface began dropping much more rapidly than that observed in the field. This drop is undoubtedly caused by the circular boundary of material in the model with a lower conductivity at 3000 ft (914 m), whereas in the field situation a tangential boundary occurs at a radius of 3000 ft (914 m). A number of alterations were attempted to improve on the simulation beyond this point. One of the things that was attempted was to change some of the elements from the 3000 to 5000-ft (914 to 1524-m) radius to the Poison Canyon and Westwater permeability rather than treating them as less permeable boundary materials; this change however produced a drawdown that was not rapid enough.

At this point another discrepancy in the data in Figs. 69 and 70 was noted. Fig. 69 shows that the peak pumping from the depressurization wells occurred sometime between 50 and 60 days after pumping starts, whereas Fig. 70 indicates that the slope of the drawdown curve was flattening out at this time. This does not seem to be a logical effect of increasing the pumping rate. Consequently, we decided to reduce the pumping rate for a few steps so as to better fit the drawdown curve. In effect, this indicates that our simulation is not correct because of the circular boundary whereas the real boundary in the field is linear. Therefore, the discharge from the depressurization wells was reduced to give a correct simulation within the ring of wells for the period to 120 days. At this time, when flow into the shaft began, the head would have to be the same as that which actually occurred. This procedure can be equated to calibrating the model by simulating the potentiometric level which occurred in the field.

Fig. 90 shows the simulated drawdown for the initial 50-day period. The agreement with the field values is quite good. As explained above, after 50 days the discharge from the depressurization wells was adjusted to give exact duplication of the potentiometric level.

The remainder of Fig. 90 is concerned with the simulation of the shaft and drift inflow during their periods of development. One difficulty with attempting to simulate the inflow to the shafts results from the grouting that was done from above the Poison Canyon Formation. The grout holes were drilled through the bottom of the shaft into the Poison Canyon before the shaft reached that formation. Inflow into the shaft occurred from these open grout holes even though the shaft had not penetrated the Poison Canyon Formation. The

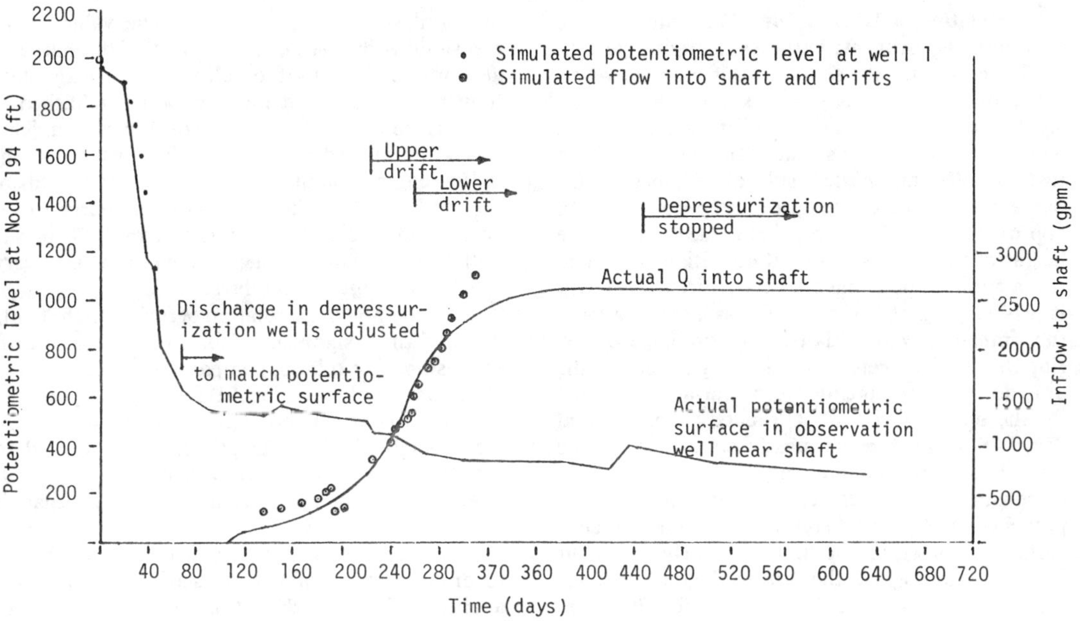

Fig. 90. Comparison of simulated and actual drawdowns and discharges.

grouting was simulated by assigning elements near the shaft a lower hydraulic conductivity than the parent material. The value of the conductivity used was approximately one-third that of the original material. This was, at best, a guess as to the effectiveness of the grouting process. The flow from these grout holes is shown by the six plotted points between 120 and 180 days. These points are slightly above the field curve; their position could have been improved by changing the conductivity of the grouted material even more.

At 180 days the shaft began to penetrate the Poison Canyon Formation. We simulated the penetration by moving the node points in the first column to a radius of 7 ft (2 m) and imposing a seepage surface boundary condition on the shaft. Our interactive computer system (Appendix D) greatly facilitated this sequence of continuous simulations. We simulated the rate of penetration of the shaft [5 ft per day (1.5 m/d)] by making runs for two-day periods at which time the next node below the bottom of the shaft was moved to the 7 ft (2 m) radius and converted to a seepage surface. Approximately 10 ft (3 m) of the shaft was open at the bottom at all times. This section was simulated by defining one node point as a seepage surface. The nodes below the shaft that still had flow because of the grout holes were allowed to continue to flow but the head was reduced successively in steps for 194 days at which time they ceased flowing. The two data points at about 200 days represent the flow into the shaft as it continued downward to the upper drift level. When the shaft reached the level of the upper drift, about 15 ft (4.6 m) of the shaft were left open to represent the inflow from the drift. Between 205 and 240 days the shaft was not being extended while work was being done on the upper drift. Subsequently the shaft commenced downward again. We continued to simulate, always keeping 10 ft (2 m) of the shaft open at the bottom by assigning one node point the seepage surface boundary condition. This process of shaft penetration continued until the shaft reached the bottom of the lower Westwater Formation.

The data points between 220 and 320 days were obtained by adding the shaft inflow to the drift inflow. The agreement between simulation and field data is excellent until 290 days at which time our simulated values continue to increase whereas the field values become constant. Tabulated data from the simulation are included in Appendix F. This difference is best explained by the method we used to calculate the drift inflow as explained in detail in the following section. For now it is sufficient to recognize that a two-dimensional analysis was used to determine the inflow into the drift per unit length of drift. Had the drifts been continued in a straight line away from the shaft this analysis would be satisfactory. However, the drifting was not done in a straight line but with curves and branches. The actual flow therefore is less than the predicted flow because the area already had been partially drained.

Simulation of Drift Inflow: We simulated the flow into the drifts with the same finite element mesh we used for the shaft. In this case, however, a two-dimensional analysis was used rather than an axisymmetric analysis. Each drift was simulated by assigning a seepage surface boundary condition to two nodes on the left side of the mesh. This procedure produced a length of seepage surface of approximately 15 ft (4.6 m) that is the approximate height of the drift. The initial condition simulated was a constant potential over the entire mesh [2210 ft (674 m)]. Using this condition as a starting base, the simulation was initiated with drainage occurring through the seepage surface. Fig. 91 shows the drawdown of the potentiometric surface with time for the upper most drift. Ultimately the potential distribution must reach steady state because the outer boundary condition for the mesh was set at a constant potential. This condition was approached at 26 days whereupon the potentiometric surface as shown on the figure is nearly a straight line that extends from the drift to the outer boundary. Fig. 92 portrays a similar curve for the lower drift but as expected a somewhat longer time was required for the steady state conditions to be achieved. The data in Fig. 93 were plotted in order to analyze the flow into the drift as a function of time after drainage started. This graph shows dimensionless discharge vs. time. The value of dimensionless discharge is the value of discharge at any time divided by the value of the steady state flow if a constant potential had been established at the surface of the Poison Canyon Formation. This was accomplished by assigning the constant potential boundary condition to all points at the tip of the mesh. Fig. 93 shows that the discharge drops very rapidly to a value of less than 10% of the initial steady state value; it then decreases very slowly to a time of 100 days.

The method of calculation of flow into the drifts required the assignment of an average rate of drift expansion. The mining company used both 10 and 50 ft per day (3 and 15 m/d) in their analyses. We used a rate of 10 ft per day (3 m/d) for our analysis. Accordingly the total length of drift required to determine the average inflow rate per foot of drift after a period of ten days after drifting began is 100 ft (30.5 m). Since part of this length had been draining for ten days whereas part had just begun to drain, a value of five days of drain time was applied to the entire drift. For the calculated values shown in Fig. 90 the upper drift was extended at 10 ft per day (3 m/d) beyond a time of 224 days. The lower drift then was extended at the rate of 10 ft per day (3 m/d) after a time of 260 days. At this time the flow into the shaft ceased because the

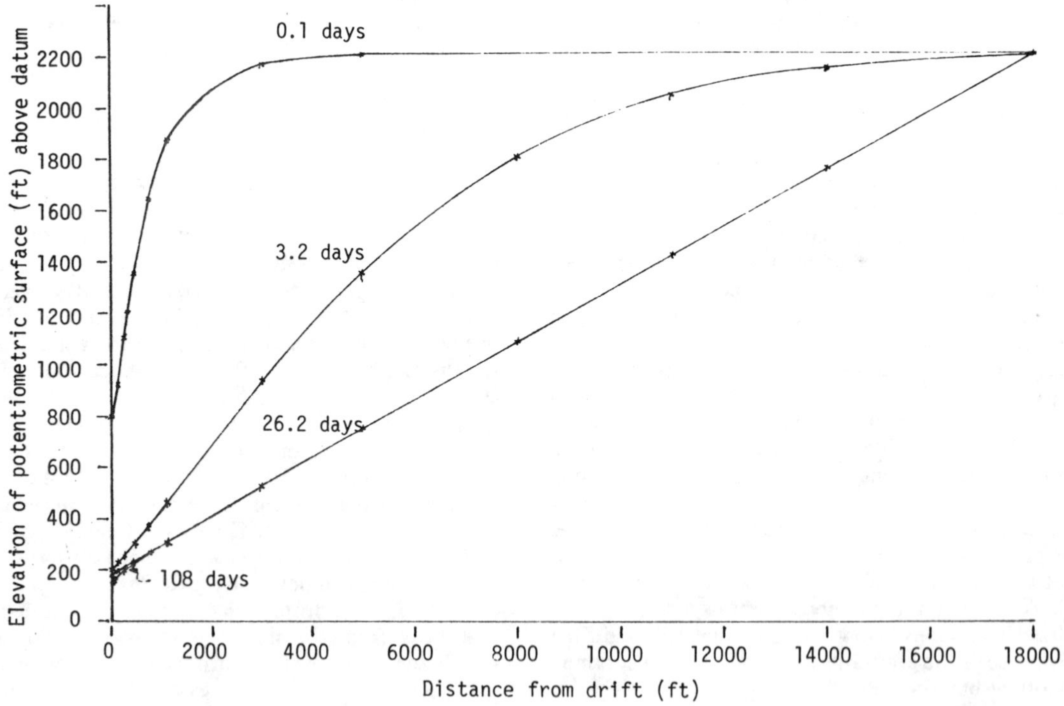

Fig. 91. Drawdown of potentiometric surface due to drift at upper level.

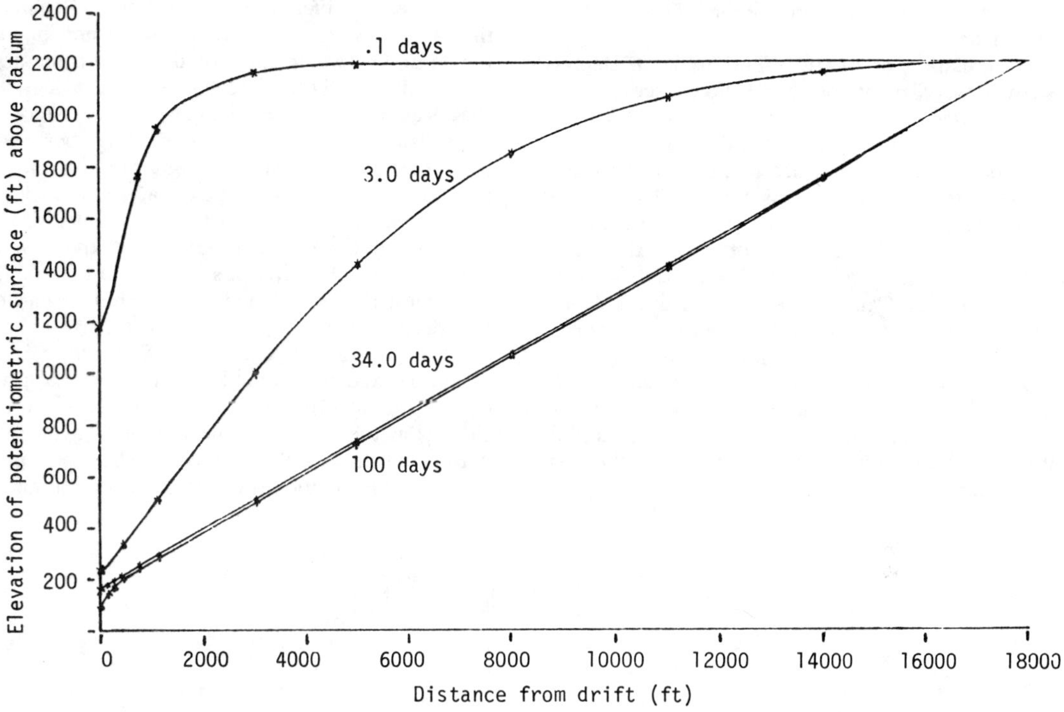

Fig. 92. Drawdown of potentiometric surface due to drift at lower level.

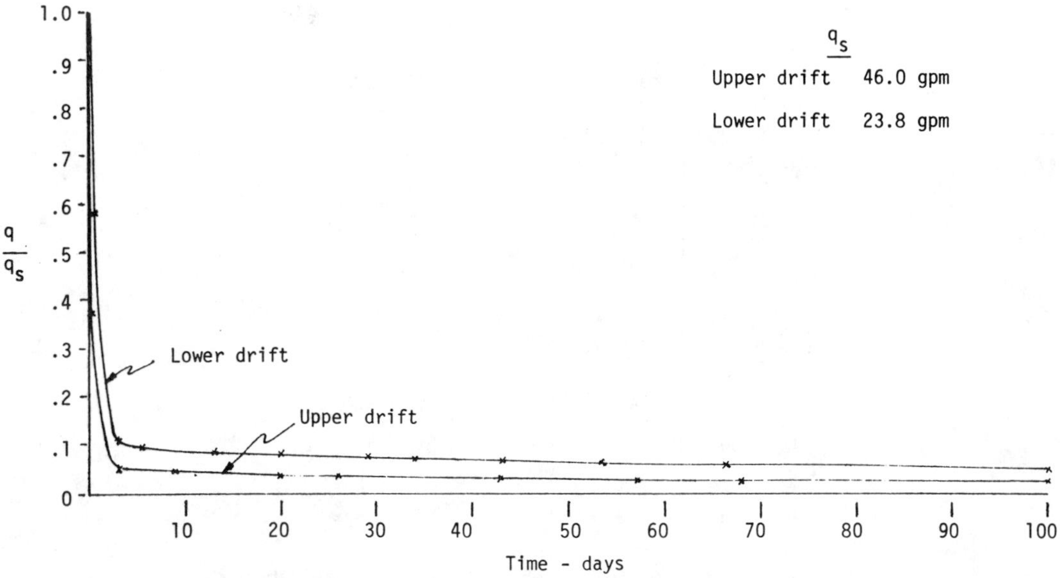

Fig. 93. Inflow per foot of drift.

bottom had been sealed and all the inflow was from the drifts.

Simulation of Shaft Inflow Without Depressurization: The data on shaft inflow discussed previously were also used in a comparison with the hypothetical flow for the hypothetical case of no boundaries, no depressurization, and no grouting. This comparison is shown in Fig. 94. The vertical coordinate is the distance from the top of the Poison Canyon Formation to the bottom of the shaft. When the bottom of the shaft was 25 ft (7.6 m) into the Poison Canyon, the inflow to the shaft was about 500 gpm (1.9 m^3/min). Had there been no boundary, depressurization nor grouting, the discharge to the shaft would have been about 4600 gpm (17.4 m^3/min). This is the type of analysis for which mathematical modeling is most useful; it shows the relative effects of various methods of operation.

These data include the flow into the shaft at the location of each drift but they do not include the additional water that would flow into the drift due to drift activities. As an example, when the shaft was at the level of the upper drift the extension of the shaft was stopped and drifting work was conducted for approximately one month. The inflow given here includes only that water that entered from the nearly 20 ft (6 m) of shaft open at the drift but not the inflow from the several hundred feet of drift. This drift inflow could be incorporated as was done in Fig. 90 but we elected to ignore it for this hypothetical case.

The general shape of the inflow curve is as would be expected on the basis of theoretical principles. The quantity of inflow is greatest when the shaft is at the middle of the Poison Canyon Formation. The inflow then decreases when the Upper Westwater Formation is exposed. This relationship

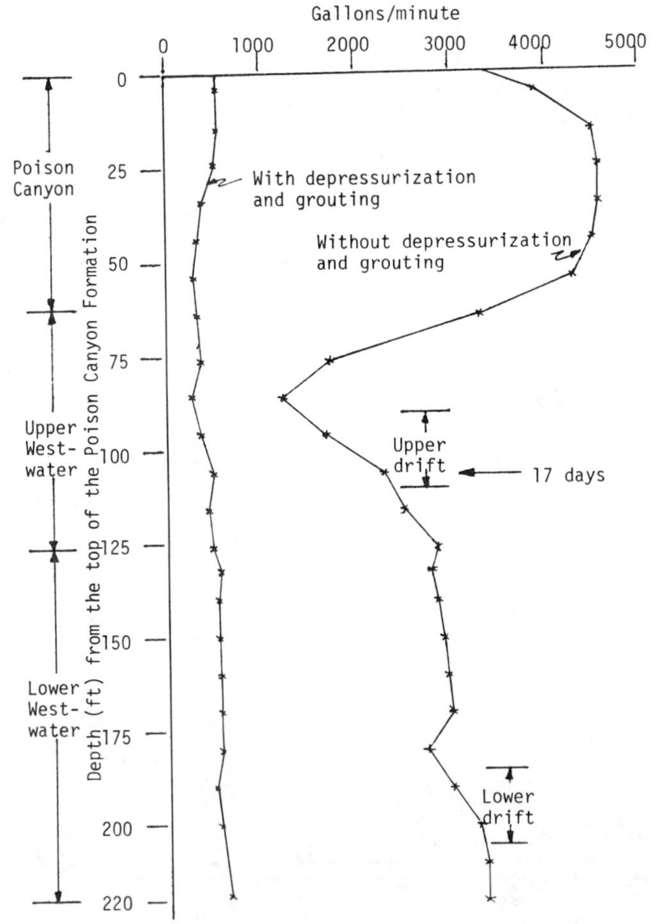

Fig. 94. Comparison of shaft inflow with and without depressurization.

is a consequence of the fact that the hydraulic conductivity of the Poison Canyon Formation is much greater than that of the Upper Westwater Formation. The inflow when the shaft is in the Lower Westwater Formation does not decrease appreciably because of the area exposed at the upper drift. Since there is more area for inflow after the upper drift is passed, the shaft inflow does not decrease to a rate equal to that immediately above the upper drift.

Summary of Uranium Mine Simulation: The simulation of the uranium mine lends considerable credibility to the methods we presented earlier in the manuscript for analyzing the flow into a shaft or a drift (or combinations) as it is being developed. Fig. 90 shows very good agreement between the simulated flow and the flow measured in the field for the period of approximately 200 days as the shaft was being sunk. The method does contain problems however for long periods of time. At later times either boundary conditions can become effective or the drifts can evolve into so many branches that the assumption of two-dimensional flow into the drift per unit width is not valid. The former problem can be minimized by more detailed hydrogeologic field mapping. As far as we know ours is the first work in which methods have been employed that allow the actual physical boundaries to change with time. As we discussed previously, the greatest problem with the axisymmetric simulation of a shaft using UNSAT2 is the question of how to simulate the effect of a tangential boundary such as a fault. Probably the most important objective to which a mathematical model such as UNSAT2 can be applied is illustrated in Fig. 94 in which two methods of operation are carried out and the consequent discharges compared.

It is doubtful that a model can be developed that is so complex that all boundaries of a developing mine can be simulated. Consequently it should be considered necessary in a mine simulation problem to attempt to determine the most important factors during a particular time period and model those important factors. As an example, the model we used in this project has limited utility for modeling the long-term effects of mining activities on the regional ground water table. By long-term effects we mean 10, 20, or 30 years. This sort of modeling is better done with a regional ground water model using two dimensions in the horizontal plane where element dimensions can be expanded and a greater area covered. In other words, it is difficult to place emphasis on modeling the inflow to a mine as it develops and to model the impact on the regional ground water flow system at the same time. Regional impact is discussed to a limited extent in the next section.

District Impact—Uranium Mine: Lowering of the regional water table or potentiometric surface is probably the most common effect of mine dewatering on regional hydrology. If producing water wells are open to a particular aquifer that is dewatered, the effect may be negative on those wells. Lowering the water table can also have an effect on surface hydrology by changing the base-flow characteristics of streams. Neither of these effects are believed to be of significance for the uranium mine we modeled. The mine is located in a very isolated area; there are no domestic or municipal wells pumping from the aquifer. In addition, the aquifer is confined and very deep and there is no good hydraulic connection between surface streams and the aquifer that is being pumped.

The discharge of mine water to streams is another possible effect on district hydrology. Such discharge could cause either improvement or degradation of water quality and change of average flows. This effect could be significant in the uranium mine basin depending on whether the pumped water is reinjected into the aquifer at some distant point or whether it is discharged into surface streams. In order to examine this effect it would be necessary to use a mass transport computer model to determine the transport characteristics of the constituent materials in the water. We did not have adequate information on the ultimate fate of all the water being withdrawn from the uranium mine to accomplish this objective. But it probably would have a significant effect, either positive or negative, on the surface water flow in the region if it were not reinjected into the aquifer.

LEAD MINE—CASE HISTORY NUMBER 2

Mine Location

The second mine we modeled is a lead mine located in the lead belt of southeastern Missouri in what is referred to as the Viburnum Trend (Fig. 95). This area lies in the Mississippi Valley, a part of the stable interior region of the United States.

Geologic Setting

Wharton (1975) summarized the geologic history and stratigraphy of the area in question as related to ore deposition. Marine deposition that began in the area in early Paleozoic time was influenced greatly by the Ozark uplift. The St. Francois Mountains, near the northeastern limit of this uplift, consist of volcanic rocks that are mainly rhyolites intruded by granite in Precambrian times. Emergence was probably as an island complex in late Cambrian seas. Deposition of the Lamotte Sandstone, the lowermost sedimentary formation, began while the uplifted area was irregular and exhibiting high relief. Fig. 96 illustrates the gen-

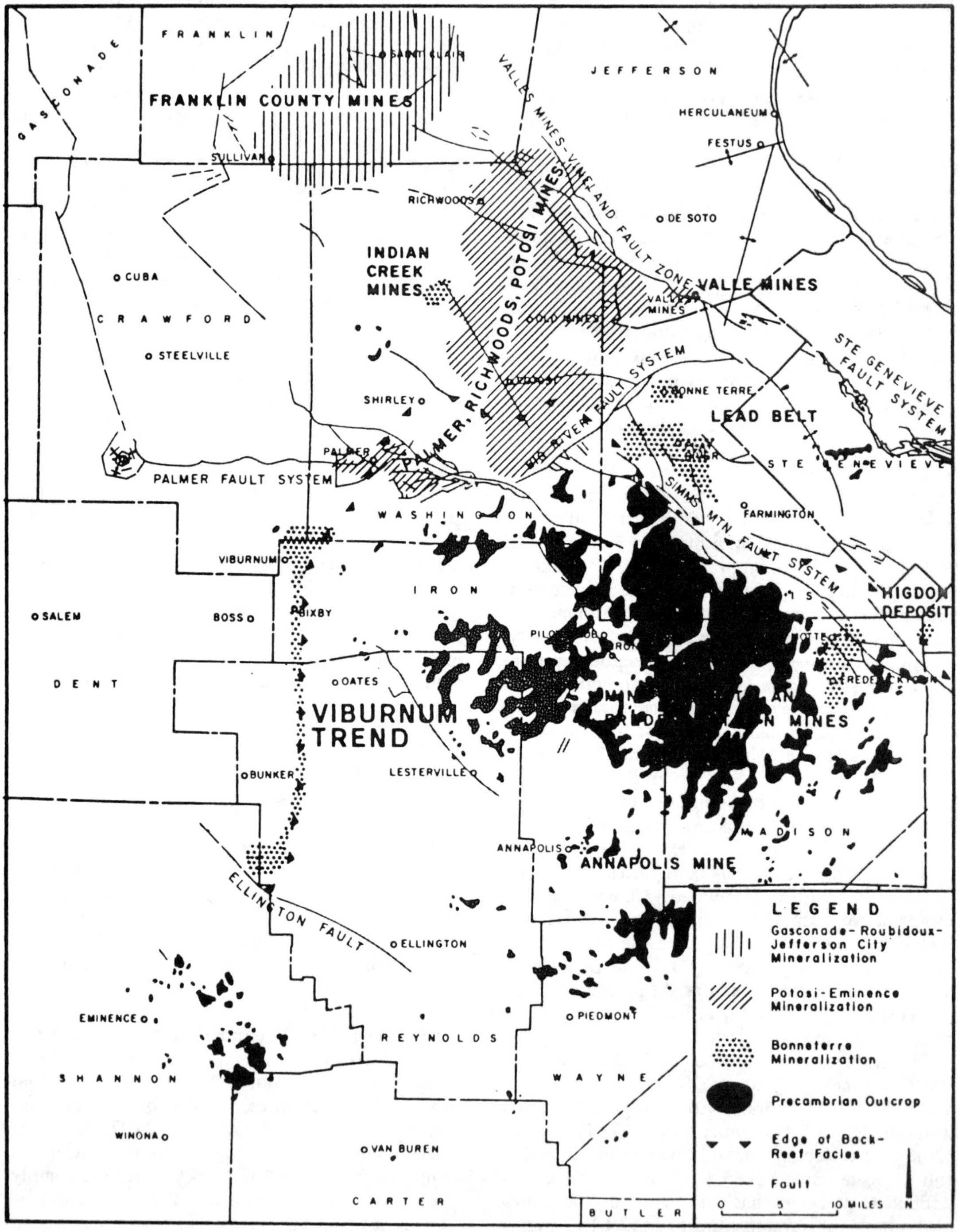

Fig. 95. Map of southeast Missouri lead district (Wharton, 1975, Fig. 1).

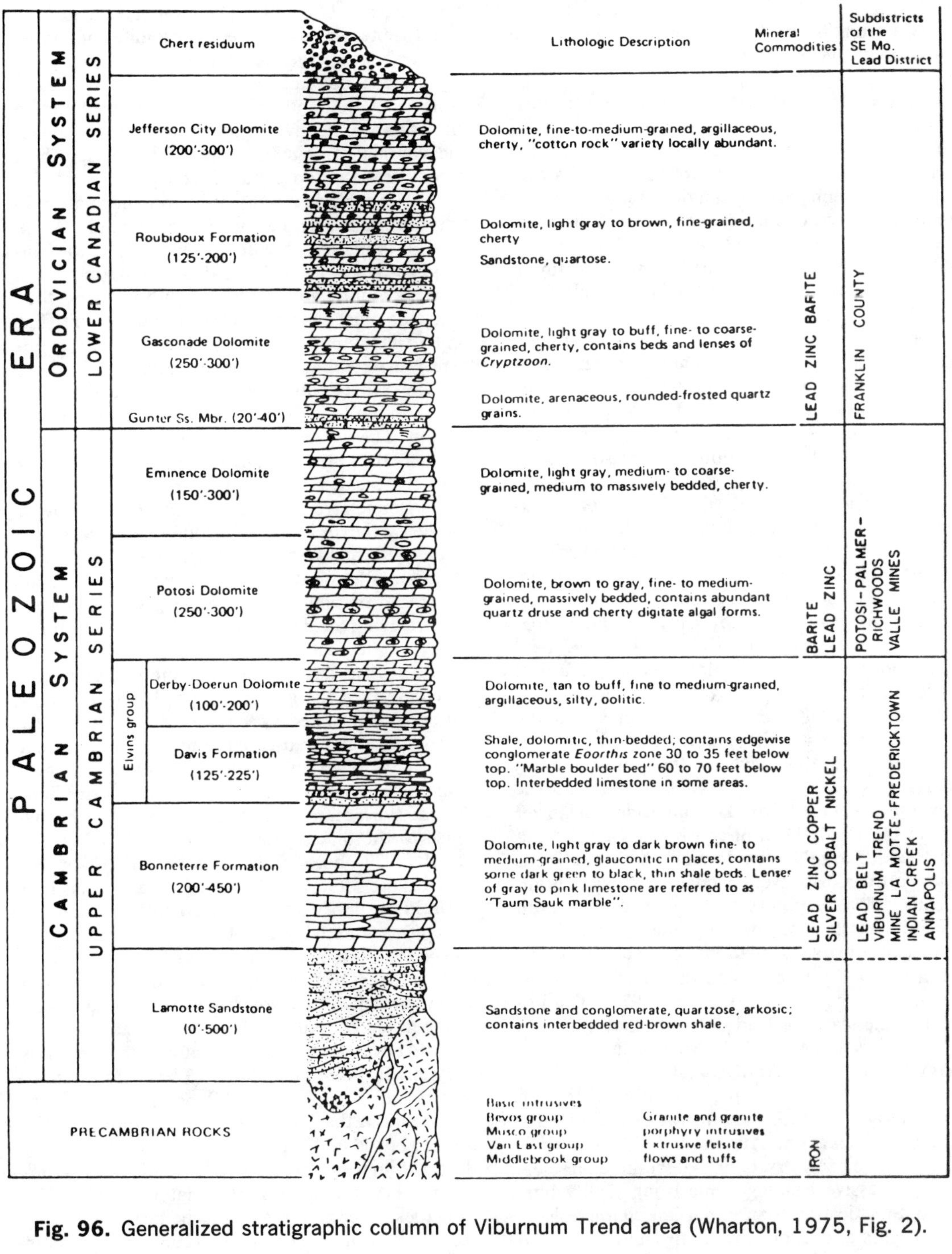

Fig. 96. Generalized stratigraphic column of Viburnum Trend area (Wharton, 1975, Fig. 2).

eralized stratigraphic column of the Viburnum Trend.

A transitional contact exists between the Lamotte Formation and the overlying Bonneterre Formation. The Bonneterre Formation consists of lime sands, algal deposits, and some lime muds. Dolomites, peripheral to the highlands, are the host rock for the most significant lead-zinc deposits in the district. The ore occurs in the vicinity of a barrier reeflike environment formed by algal structures in the Bonneterre Formation. Limestone predominates over dolomite in an offshore direction. Three known explosive volcanic episodes occurred in Late Cambrian times (early Bonneterre time). Two of the three occurrences are known only from core data.

The contact between the Bonneterre Formation and the overlying Davis Formation appears to be unconformable throughout most of southeast Missouri according to Thacker and Anderson (1977) but it is conformable in some areas. Thacker describes the Davis Formation as consisting commonly of interbedded green shales, calcareous siltstones, and sandy and silty limestones. Mudchip conglomerates occur commonly in the Davis. Wharton (1975) noted that the Davis shale has been considered an impermeable barrier to ascending ore genesis theories or a semipermeable barrier in descending ore genesis theories. This hydraulic property of the Davis is critical to the successful completion of our mathematical modeling effort at the lead mine.

Thacker and Anderson (1977) notes that the Derby-Doerun Dolomite conformably overlies the Davis Formation everywhere; the contact occurs where the green shales of the Davis end and the brown shales of the Derby-Doerun begin. The lower unit of the Derby-Doerun consists of thin, irregularly bedded argillaceous dolomite with numerous shale partings. Massively bedded, burrowed, argillaceous dolomite occurs in the upper unit. Both units are fine to medium-crystalline in character.

A conformable, gradational contact occurs between the Derby-Doerun and the overlying Potosi Dolomite (Thacker and Anderson, 1977). The Potosi is a massively bedded dolomite that is fine to medium-crystalline, highly burrowed, and typically vuggy (Thacker and Anderson, 1977).

The Eminence Dolomite is conformable with the underlying Potosi; the interval between is gradational (Thacker and Anderson, 1977). The Eminence is a coarsely crystalline dolomite characterized by massive bedding while being highly burrowed. Silicification is an important characteristic; silica occurs also as massive fillings and interstitial fillings between dolomite crystals. Quartz sand is common in the upper Eminence.

Warner et al. (1974, pp. 38–40) summarizes the characteristics of the Gasconade Formation and its basal member, the Gunter Sandstone. The Gunter Sandstone was deposited on an erosional surface of the Eminence Dolomite. The Gunter Sandstone is usually noted as an arenaceous dolomite containing thin, discontinuous sandstone lenses. Light gray, argillaceous dolomite overlies the Gunter Sandstone. This dolomite grades into a gray dolomite that is more massive, medium to coarse-grained, and containing abundant chert. The denser chert occurs near the top of the Gasconade.

The Roubidoux is the uppermost formation in the Viburnum Trend area. Warner et al. (1974, pp. 40–41) summarizes the descriptions of previous authors to describe the Roubidoux as a complex of chert, dolomite, and sandstone with the sandstone being the most conspicuous stratigraphic unit. The thin to medium bedded sandstone exhibits abundant cross-bedding and ripple marks.

Mineralization in the Viburnum Trend usually occurs in the upper sections of the Bonneterre Formation. The general consensus concerning the origin of the ore assumes that the ore-bearing solutions created the collapse breccia that occurs in the Bonneterre. The path the solutions followed and hence the location of the collapsed breccia were predetermined by structural controls and the distribution of hydraulic conductivity within the Bonneterre Formation. The ore-bearing solutions were the dissolving media that followed the path of highest hydraulic conductivity.

Geologists for several area mining companies have described the geologic controls on ore distribution at their mines. Sweeney, et al. (1975) indicate that solution-induced collapse breccias are evident in all areas of extensive mineralization. The mineralization is restricted almost totally to the Bonneterre Formation; the relationship of ore occurrence to the breccia zone is indicated in Fig. 97. Davis (1977) notes that the most important ore controls are solution-induced collapse breccias that occur in the upper 100 ft (30.5 m) of the Bonneterre Formation. Paarlberg and Evans (1975) notes that galena (the principal sulfide ore) fills open spaces such as fractures; solution collapse breccia and slump breccia often contain mineralization. The better grade of mineralization occurs in the Bonneterre Formation. The mineralized zone extends occasionally into the lower Davis Formation. Mouat and Clendenin (1975) notes that fracture ore occurs in the Bonneterre Formation below the breccia but is not important economically. One of the most important ore bodies is bedded or disseminated where the ore has spread laterally within the host rock. Breccia ore tends to be disseminated or concentrated in vugs while marginal ore occurs in a complex system of faults or breaks parallel to the breccia margin. Fig. 98 illustrates the various types of ore concentrations noted by Mouat and Clendenin (1975).

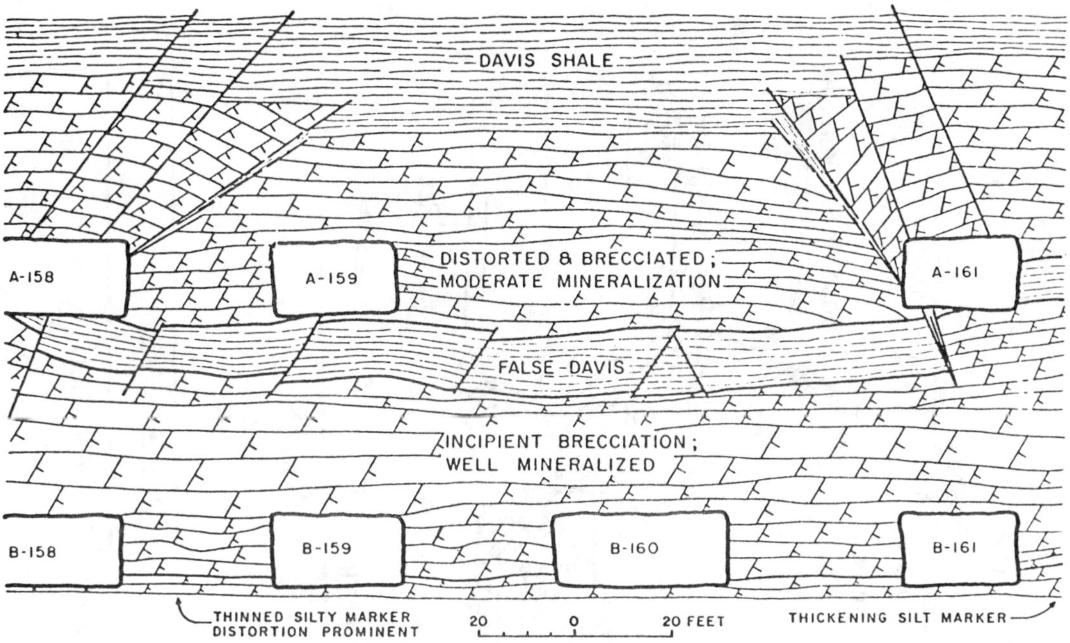

Fig. 97. Relationship of ore occurrence to breccia zone and False Davis (Sweeney, et al., 1975, Fig. 4).

Not all ore occurs in or is associated with the brecciated zone. Grundmann (1977) notes that the mineralization occurs in calcarenite-filled channels between linear algal mounds and troughs that were formed behind subtidal offshore algal bars.

Mine History

Underground development of the lead mine began in May 1964 with the drilling of a 7-ft (2 m) diam shaft that is 1250 ft (381 m) deep. Two mine levels were developed from this shaft and subsequently intersected by a second, 20-ft (6 m) diam, concrete-lined production shaft in April of 1967. Mine production began in June of 1968.

The multilevel ore body is mined using the open stope room-and-pillar method. Ore is moved through raises between levels to a diesel rail haulage system on a sublevel. Ore is transported by rail to ore pockets and an underground primary jaw crusher. The ore is hoisted automatically in two skips.

Mine water inflow does not exist in any significant amounts at this mine except where the breccia zone is intercepted within the Bonneterre Formation. Water enters the mine through cracks and vugs in the back (roof), walls, and floor of the drifts. The water-producing zone is bounded by the dimensions of the mineralization zone in the breccia. The rate of water inflow in a drift decreases with time from the opening of the drift but the overall mine flow rate remains essentially constant with time because of the expanding drifts. Inflow is estimated to be 600 to 800 gpm (2.3 to 3 m^3/min) without any noticeable seasonal increase or decrease in inflow. We interpret the absence of fluctuations to be indicative of no interconnection between the ground water and surface water systems. Inflow essentially has been constant without any sudden large inflows of water although a fault intercepted by one drift was and is a source of inflow.

Data Collection

Surface Water System: Some surface water data were collected at the mine site. Precipitation data and streamflow water quality data were collected for the area surrounding the mine. These data did not prove very useful to our modeling effort because of the presence of the Davis shale above the Bonneterre Formation.

Ground Water System: Ground water inflow into the lead mine varies between 600 and 800 gpm (2.3 to 3 m^3/min). These figures must be considered as estimates because continuous measurements are not available at this site.

Extensive potentiometric level data exist at this lead mine. However, there are no continuous well hydrographs from the inception of mining to the current data because as old wells are destroyed by mining new monitor wells have replaced them. The earliest data available on potentiometric levels begin in June 1965. However, that well was mined

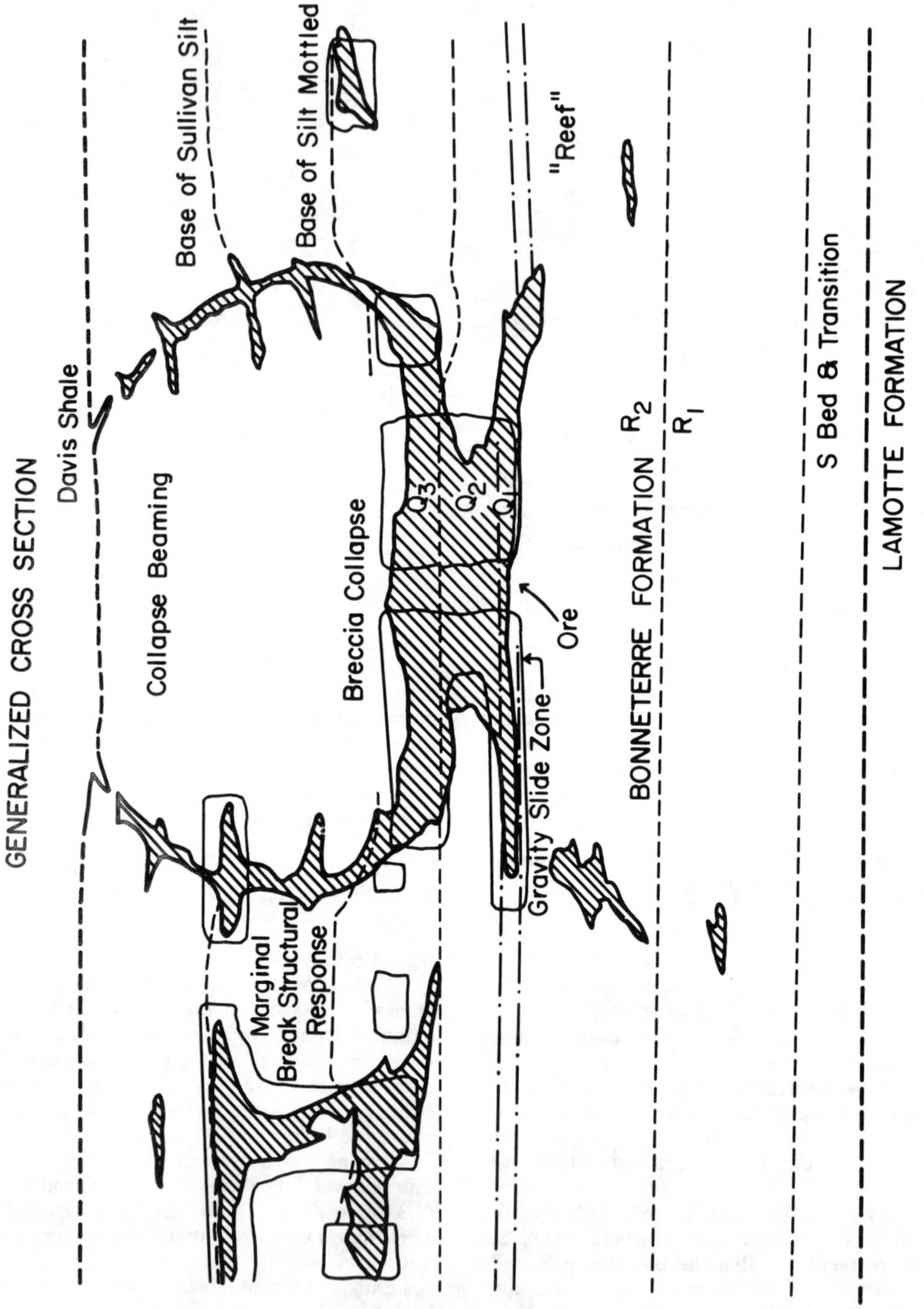

Fig. 98. Relationship of ore occurrence to breccia zone, cross section (supplied by the mining company).

out in June 1970. Three additional wells were selected that have a continuous record of water levels from as early as January 1970 until August 1980. These wells in combination give a continuous water level record from as early as June 1965 through the present. We obtained these data as a mixture of prepared graphs and tabulated water level measurement data.

An estimate of premining water levels can be obtained from the static water level data taken prior to pump tests conducted in the area. The four observation wells addressed previously are shown in Fig. 99. All four wells are open only within the Bonneterre Formation. A limited water level record also is available for the Potosi Formation. The location of this well is shown in Fig. 99 also. Water level data for this well began in April 1979 and extended through November 1979. There were no noticeable fluctuations of the water levels measured in the Potosi Formation during this period. This flat water level trend indicates a lack of hydraulic connection between the Potosi Formation and the Bonneterre Formation since nearby observation wells in the Bonneterre Formation indicate a steady decline in water level within the Bonneterre Formation due to mine pumpage. Water level data for these same observation wells in the Bonneterre Formation also indicate a lack of hydraulic commu-

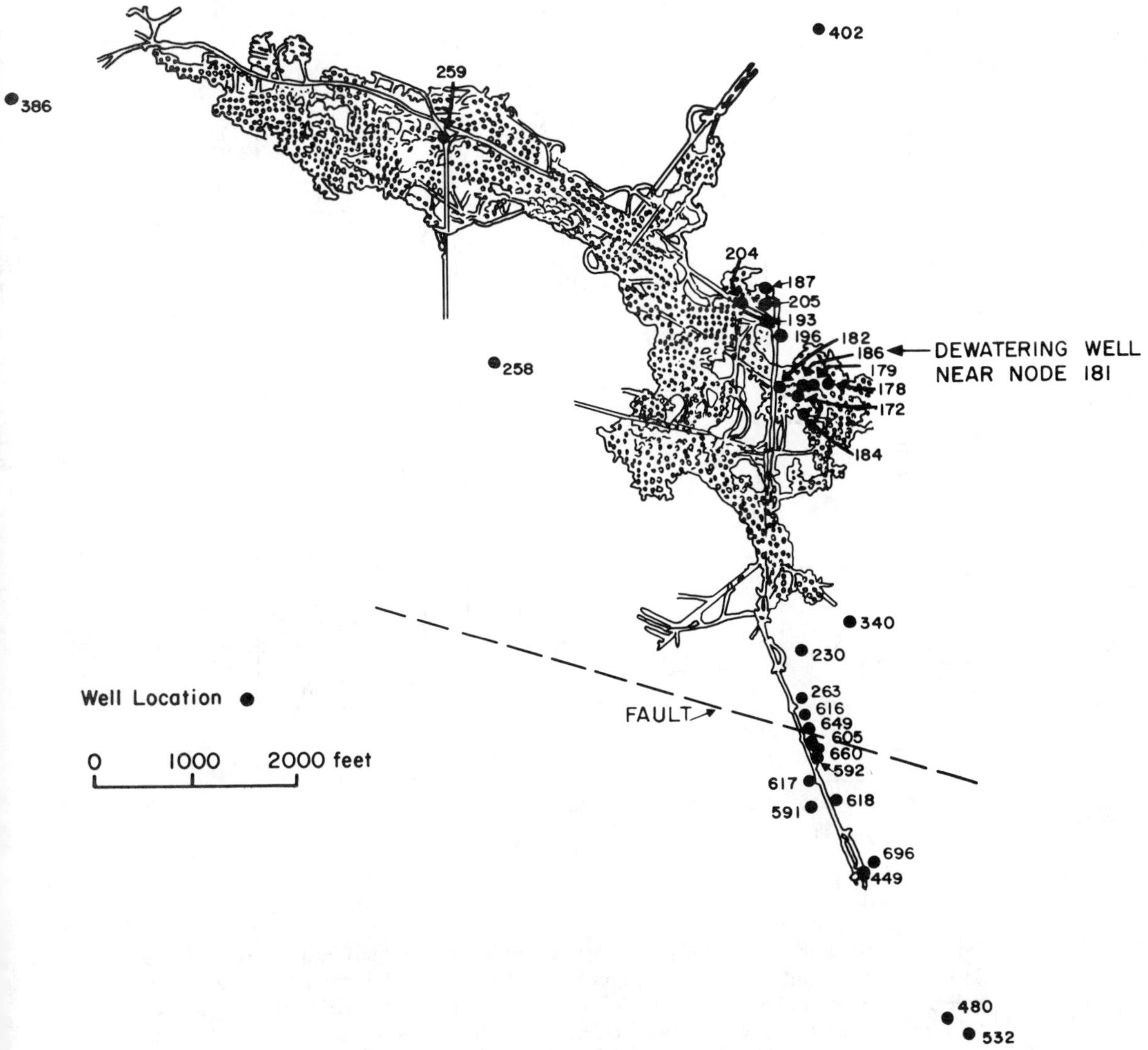

Fig. 99. Well locations, lead mine (supplied by the mining company).

nication across the fault shown in Fig. 99. Pump downtime is reflected in the pumping well as well as in the two observation wells located on the same side of the fault. The one observation well shown on the opposite side of the fault from the pumping well does not reflect any water level change due to this pump shut-down. Fig. 100 is a copy of water level data obtained from the lead mine that illustrates this point. We consider these water level data to be evidence that the Potosi Formation can be excluded from consideration of mine inflow. These data also indicate that the fault can be considered as a barrier to ground water flow on one side of the mine. Fig. 101 is a graphic representation of the ability of the fault to act as a barrier to ground water flow.

Numerous pump tests also were conducted in the vicinity of the lead mine. The initial pump tests in the area were run in 1961, 1963, and 1964. Pump tests were conducted also in 1975 and 1979. The initial pump tests in 1961 and 1963 are outside the area of interest with respect to the objective of this manuscript.

The pump tests conducted in 1964 included both the Bonneterre and the Potosi Formations. These tests were conducted by the mining company

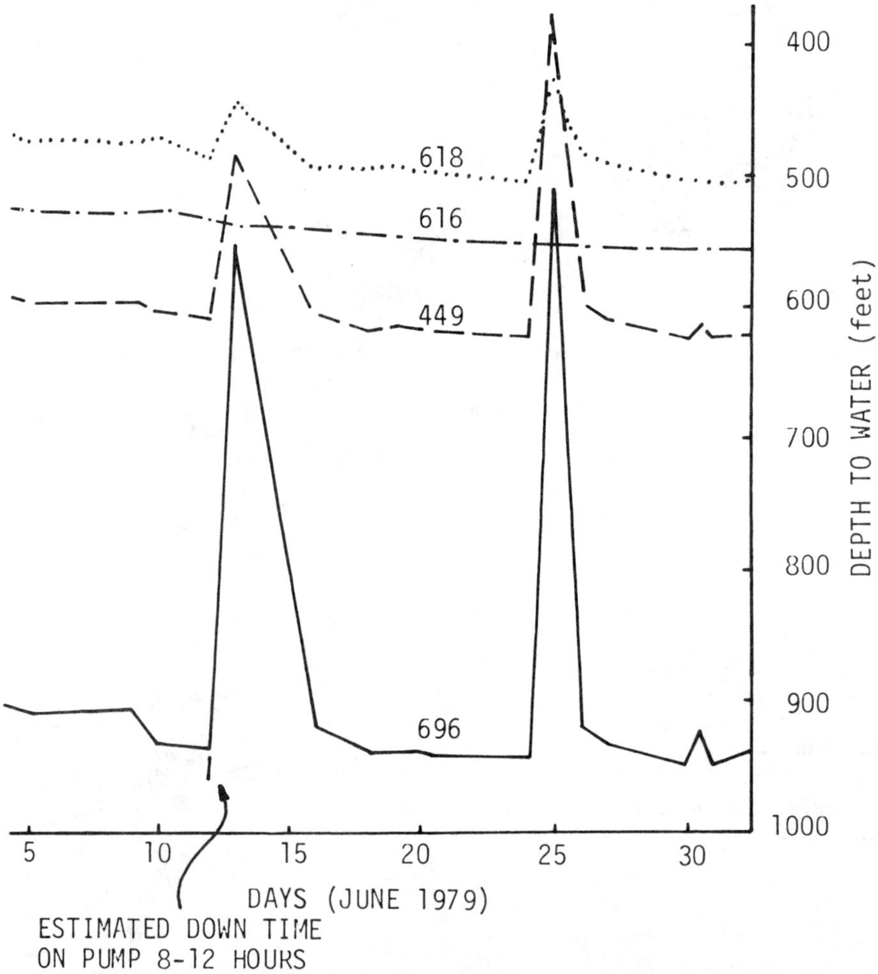

Fig. 100. Ground water levels in the vicinity of the lead mine (Bonneterre Formation, supplied by the mining company). Note: Potosi Formation observation well 616 had no noticeable change in water level for an 8-month period of record, although observation wells 449, 616, and 696 showed continuous decline. Observation wells 449, 618, and 696 are on the same side of the fault while 616 is on the other side of the fault.

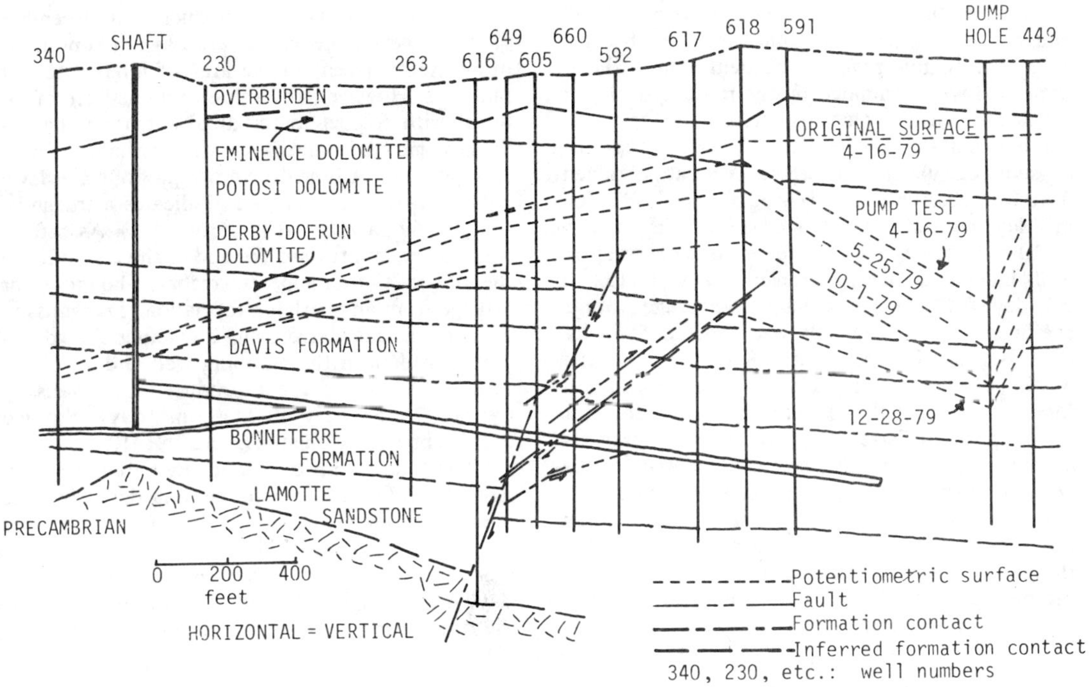

Fig. 101. Effect of fault on the ground water levels in Bonneterre Formation (supplied by the mining company).

in order to predict dewatering requirements at the exploration shaft site at this mine. The Bonneterre Formation pump test consisted of a 24-hr continuous rate pumping test with data collected both within the pumping well and from two observation wells. The locations of these wells are shown in Fig. 99. Continuous pumping from the Bonneterre Formation began 12 days after the conclusion of the controlled pumping test.

Observations from both pumping periods apparently were used by the mining company to evaluate dewatering requirements. The average transmissivity calculated from water level data taken from the observation wells is approximately 750 gal/day/ft (9.3 m^2/day) and the storage coefficient is approximately 3×10^{-4}. These data were used to predict drawdown conditions after 30 days of continuous pumping at rates of 350 and 400 gpm (1.3 and 1.5 m^3/min). Drawdown predictions were based also on 100 days of continuous pumping. Neither set of drawdown predictions allows for appreciable leakage from the Potosi Formation.

Water level data taken during long-term pumping for dewatering purposes indicate the presence of boundaries in the vicinity of the mine (Fig. 102). The initial boundary was encountered after the first day of pumping and the second boundary was encountered on the fifth day of pumping. Both boundaries are indicated by a steepening of the drawdown data curve as plotted on the semilog plot. Drawdown data gradually flattened until the 12th day. The subsequent gradual flattening of the drawdown trend shown in Fig. 102 could be due to leakage from the overlying formation or it could be caused by the intersection of an unknown recharge boundary. The mine hydrologist attributed it to leakage and estimated a rate of 400 gpm (1.5 m^3/min) continuous pumping rate. It was concluded that pumping at the exploration shaft site would not exceed 1000 gpm (3.8 m^3/min) and, in all probability, a maximum rate would be nearer 600 to 700 gpm (2.3 to 2.6 m^3/min).

However the mine hydrologist subsequently decided, apparently on account of the uncertainty, to make a second evaluation of the same pump test data (Fig. 103) using different storage coefficients. The results differed significantly from the original evaluation. Transmissivities were found to range between 1000 and 1400 gal/day/ft (12 and 17 m^2/day) per ft when storage coefficients ranged between 0.6 and 0.05. In order to address better the question of the role of the Potosi Formation a 24-hr pump test was attempted on the Potosi Formation in October of 1964. However, power supply and other problems forced the test to be abandoned after a short period of pumping. Finally a pump test was conducted with a reduced time of only 82 min of operation; at that time the pump became plugged with fine grained sediment. These short-term data were evaluated by the mine hydrologist via drawdown and recovery techniques for transmissivity and storage coefficient. The average transmissivity calculated from water level data obtained from four observation wells is 5300 gal/day/ft (66 m^2/day) with an average storage coefficient of 6 $\times 10^{-3}$. Recovery data for these same observation wells indicate an average transmissivity of 115,000 gal/day/ft (1400 m^2/day) and an average storage coefficient of 4 $\times 10^{-3}$. It was concluded finally that four to six 5000-gpm (18.9-m^3/min) wells would be required to dewater the Potosi Formation for the sinking of the exploratory shaft. However the number and size of the wells, the power requirements, and the high sediment content of the water that would be pumped all indicated that a dewatering scheme in the Potosi Formation was not feasible. Consequently it was recommended that other methods such as grouting and/or freezing be used to facilitate sinking the shaft through the Potosi Formation.

A pump test also was conducted in December 1975. However, this pump test was not conducted in an area adjacent to the area of interest to our purposes. However, the values obtained are of interest with respect to the areal extent of the hydrogeologic environment and the possibility of data extrapolation. A time drawdown plot of water level data from the pumping well indicates a transmissivity of 69 gal/day/ft (0.86 m^2/day). As before, water level data toward the end of the test indicate the influence of an apparent recharge boundary or leakage from an overlying formation. Transmissivity also was calculated using a distance drawdown plot with data in the pumping well and an observation well; this approach produced a transmissivity of 275 gal/day/ft (3.41 m^2/day). Storage coefficient was calculated to be 2 $\times 10^{-5}$.

The final pump test conducted took place in April 1979. Drawdown and recovery data from this pump test were evaluated for the pumped well and one observation well. The transmissivity for the pumping well appears to be approximately 40 gal/day/ft (0.50 m^2/day), while the transmissivity for the observation well appears to be approximately

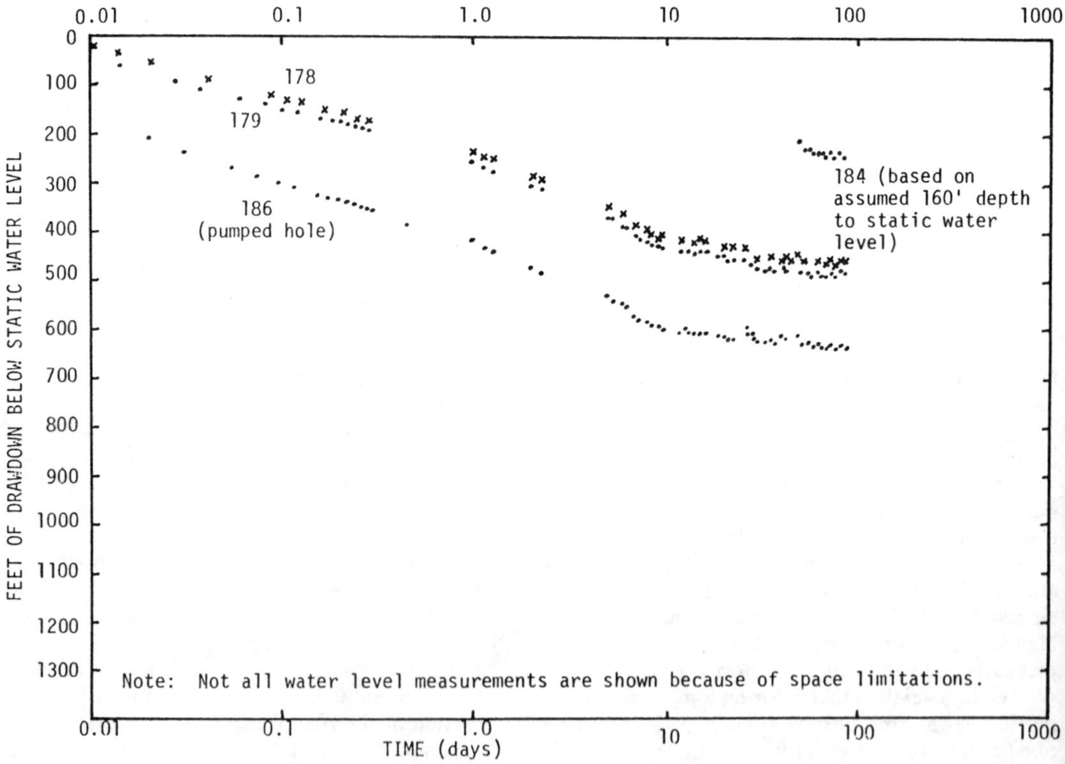

Fig. 102. Pump test data indicating presence of boundaries in Bonneterre Formation, 1964 (supplied by the mining company).

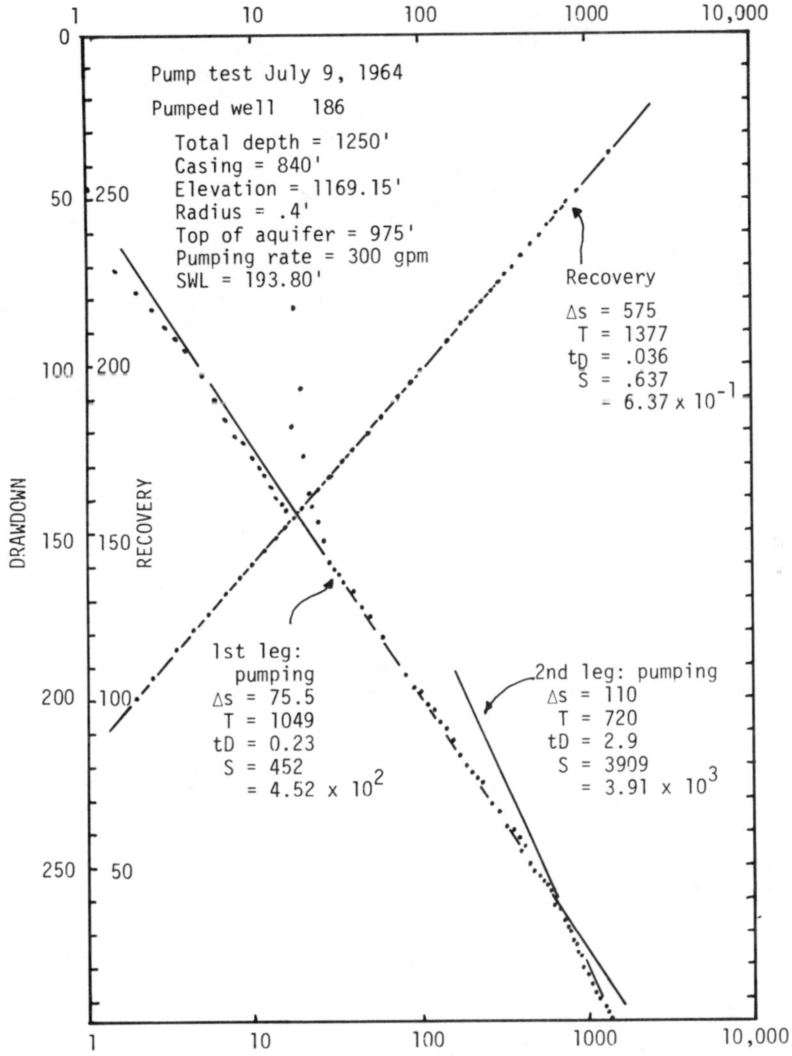

Fig. 103. Bonneterre Formation pump test, 1964 (supplied by the mining company).

45 gal/day/ft (0.56 m^2/day). The storage coefficient at the observation well was calculated to be 2×10^{-4}. These data apparently were reviewed a second time; however, the second evaluation also used drawdown data from an additional observation well (Fig. 104). A transmissivity of 47 gal/day/ft (0.58 m^2/day) was calculated in the second evaluation as compared to the previously calculated 45 gal/day/ft (0.56 m^2/day). The second storage coefficient is approximately 1×10^{-5} vs. the original 2×10^{-4}. Transmissivity calculated for the additional observation well is 146 gal/day/ft (1.81 m^2/day) with a storage coefficient of 7×10^{-6}. Drawdown plots of the two observation wells appear to reflect dissimilar boundary conditions. Drawdown data from the first observation well indicate a recharging boundary approximately 210 ft (64 m) from the pumping well. A second observation well, however, shows a discharge boundary effect approximately 830 ft (253 m) from the pumping well (Fig. 105).

Only the pump test conducted in 1964 was conducted within the breccia zone of the Bonneterre Formation; all other pump tests were conducted in the nonbrecciated zone. The difference in transmissivity of the brecciated vs. the nonbrecciated Bonneterre is readily apparent when the values of transmissivity from the two zones are compared.

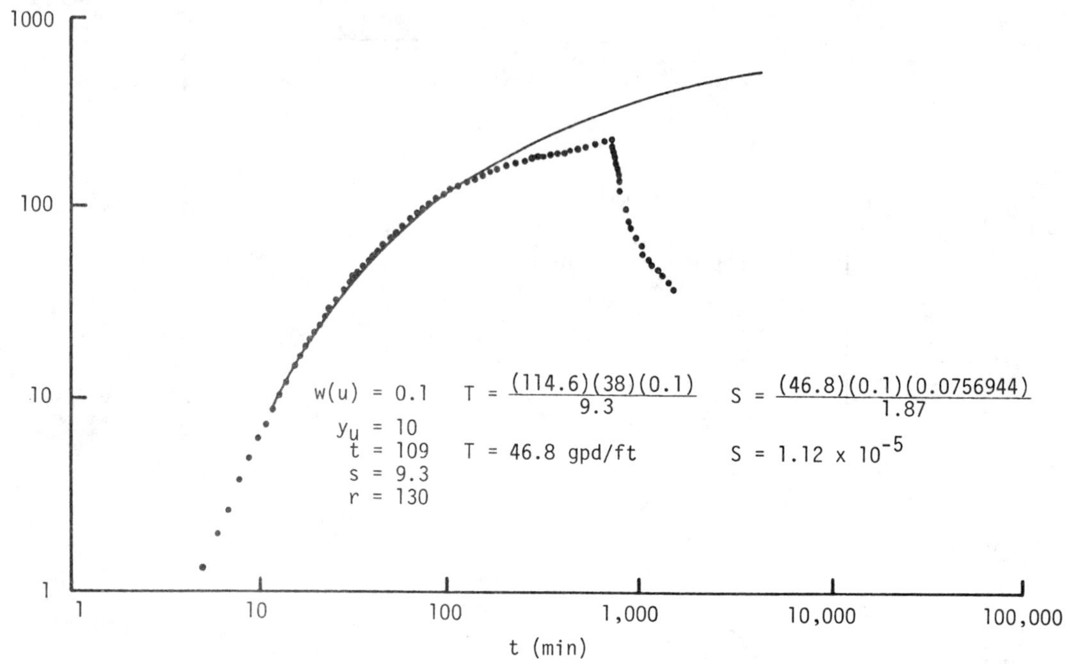

Fig. 104. Bonneterre Formation pump test analysis, 1979 (supplied by the mining company).

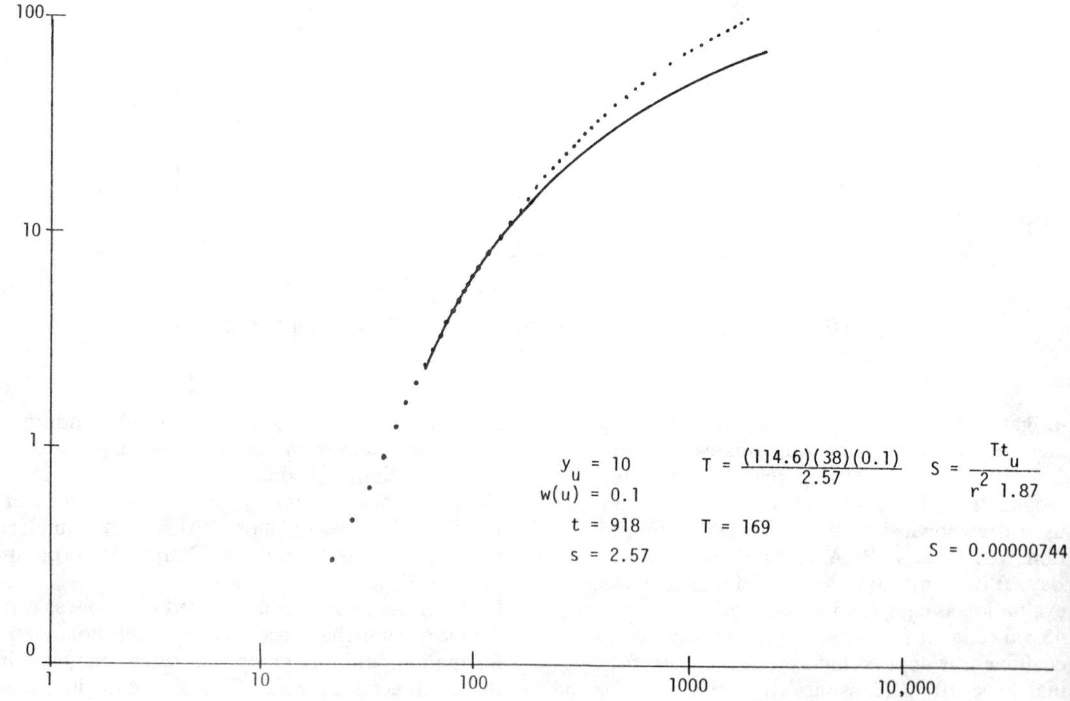

Fig. 105. Bonneterre Formation pump test indicating discharge boundary, 1979 (supplied by the mining company).

Transmissivity within the brecciated zone is approximately an order of magnitude higher than in the nonbrecciated zone. This example illustrates very well the necessity for identifying the hydrostratigraphic units in the vicinity of a mine and of monitoring and testing them individually.

On the basis of these tests we concluded that the hydrogeologic regimen at the lead mine can be described adequately in terms of transmissivity and storage coefficient determined from the on-site pump tests and from water level trend data. We concluded that the Potosi Formation that overlies the mine host rock apparently does not yield significant amounts of water to the mine because of the presence of the Davis shale that lies between them. In addition the fault shown in Fig. 99 apparently acts as a boundary to flow within the Bonneterre Formation. This fault appears to act as a recharging boundary from the side of the fault opposite to that in which most of the mining is now occurring. Mining occurs on the north side of the fault. That section of the mine suitable for our model is bounded to the south by this fault, to the north by the breccia boundary, to the east by the edge of the breccia, and to the west by a facies change. Surface water systems are unaffected by discharge into the mine from the Bonneterre Formation because of the Potosi Formation. However, the near surface formations and streams are recharged by discharge from the mine under an NPDES (National Pollutant Discharge Elimination System) permit.

Simulation of Lead Mine Inflow

We simulated the lead mine in a horizontal plane rather than in a vertical plane or axisymmetricly as we treated the uranium mine. This decision was precipitated by the fact that the mine is located in a complex hydrogeologic environment that exhibits four boundary conditions; in addition the mine utilizes a room-and-pillar extraction procedure as opposed to simple horizontal drifts. As far as we know our method of simulating room-and-pillar mining by "mining out" elements is unique. It is described in greater detail in the following sections. At this point let it suffice to say that the technique is very effective.

In order to use UNSAT2 for horizontal plane flow simulation it is necessary to assume that the thickness of the formation is uniform and that the mined out region extends over the entire vertical dimension of the formation. This means that a constant transmissivity is assumed; in reality in the region immediately adjacent to the mine the transmissivity would actually be reduced as drawdown develops. After considerable experimentation we concluded that the effect on the rate of mine water inflow was small and could be safely neglected. Fig. 106 shows the finite element mesh that we used for the simulation. It consists of 258 node points and 281 elements. The ore body to be mined is shown as the shaded portion of the figure. Table 10 gives the values of material properties that we assigned to the mesh. We considered these to be the best values that could be derived from the pump test data.

In the process of simulation we initially pumped node 181 at the location of the dewatering well (see Fig. 99). The initial conditions for the potentiometric surface over the entire flow field were taken from field data that give the potentiometric surface in 1964 before any pumping began. The boundary conditions employed consisted of holding the top, right, and left sides of Fig. 106 at a constant potential. The bottom of the mesh is bounded by the fault zone across which we allowed no flow. These boundary conditions were derived from the best field data available.

Node 181 was pumped at a rate of 400 gpm (1.5 m^3/min) for 273 days at the start of simulation. This rate and duration of pumping were based on information provided by the mining company about the dewatering program. During this stage nearly 600 ft (183 m) of drawdown developed at the pumped well. The next stage of simulation consisted of pumping node 181 at a rate of 850 gpm (3.2 m^3/min) for a period of 30 days. This was the time at which production began in 1965. No data are available on the rate of expansion of the mine so we were forced to assume that the rate was relatively constant between 1965 and 1980. In order to simulate the mining process, an element periodically would be changed to a very high hydraulic conductivity material in order to represent a mined-out portion through which there would be no loss of potential. The potential at the boundary of the mined out portion was also set at zero at this time so as to represent atmospheric pressure. Element 189 was the first element to be mined out. This development required nodes 160, 161, and 181 to be set at zero potential. Simulation occurred for a period of 305 days during which time the inflow averaged about 700 gpm (2.6 m^3/min); it decreased from 1400 gpm (5.3 m^3/min) initially to 575 gpm (2.2 m^3/min) at the end of the 305 days. At this time element 167 was mined out and nodes 142 and 143 were changed to a zero potential in order to simulate the development. This caused an initial increase in flow to nearly 1500 gpm (5.7 m^3/min) that then dropped rapidly to 800 gpm (3 m^3/min).

This procedure was followed for a total period of 5843 days (16 years) at which time the mined-out elements in the ore body (and aquifer) were the same as the mined-out portion of the room-and-pillar mine in 1980. During this period the inflow to the mine dropped to as low as 419 gpm

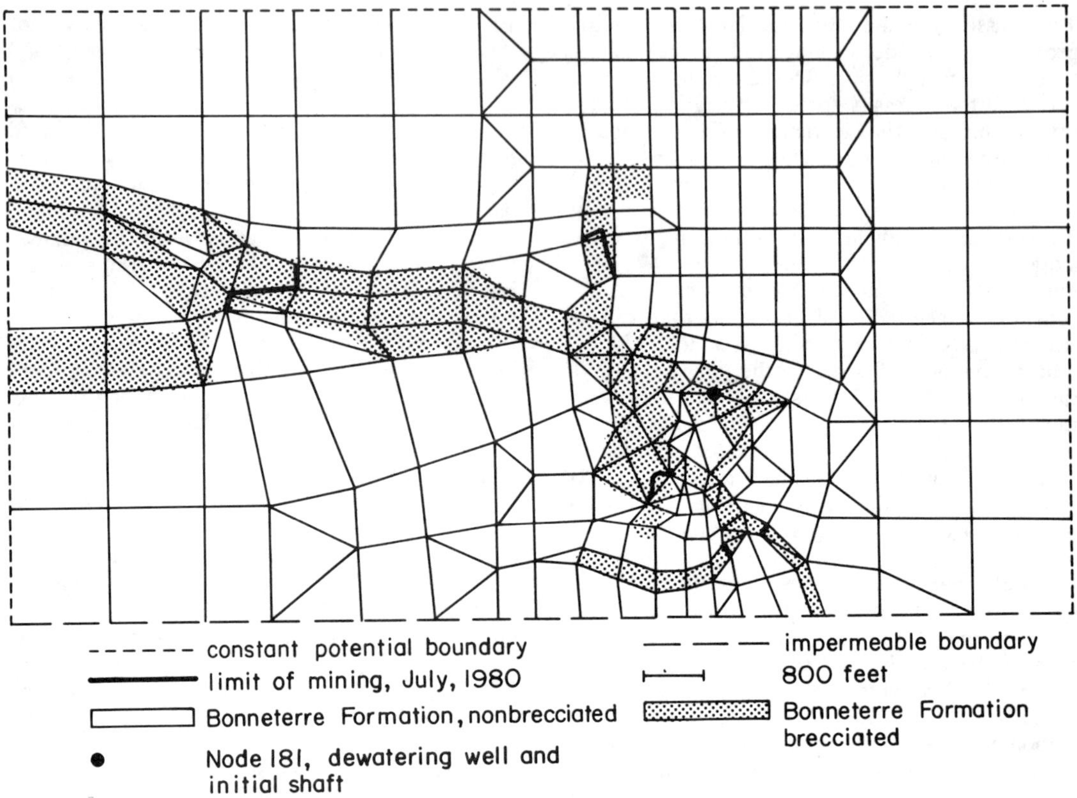

Fig. 106. Finite element mesh for the lead mine.

Table 10. Values of Material Properties Used for Lead Mine Simulation, Bonneterre Formation

Saturation, S	Moisture content, θ	Capillary pressure, ft of water, ψ	Relative conductivity, K_r	Saturated conductivity, sq ft per day, K
		Breccia, Mined		
0.99	0.99	10	0.99	1×10^{20}
1.00	1.00	0	1.00	
		Breccia, Unmined		
3×10^{-3}	0.0001	10	1×10^{-6}	100
1.00	0.3	0	1.00	
		Nonbreccia, Unmined		
3×10^{-3}	0.0001	10	1×10^{-6}	10.7
1.00	0.3	0	1.00	

Note: Saturated conductivity is transmissivity in table since model uses a unit thickness of hydrostratigraphic unit when used in a horizontal mode. Only two points for the saturation-capillary pressure-relative conductivity relationships were used because the model does not allow desaturation to occur in horizontal plane flow and the values are not used in the actual computation. However, at least two values are necessary for the program to run.

(1.6 m^3/min) after nearly ten years of operation. It then increased and remained in the range of 600 to 900 gpm (2.3 to 3.4 m^3/min) for the remainder of the 16 years. This variation is more than that reported by the mining company but the difference is explainable by the order in which the elements were removed. The mine operators reported a maximum inflow of about 800 gpm (3 m^3/min). The procedure we used for simulation was to mine out the area around the shaft completely and then proceed toward the left side of Fig. 106. For simulation this was the simplest approach possible. We elected to use the simplest approach because accurate information on exactly how mine expansion did take place was not available in the mining records. The simulated values of pumpage and potentiometric surface for selected time steps are given in Table 11, Fig. 107, and Appendix F.

When using this method of "mining out" elements, several factors should be noted with respect to simulation results and discrepancies that may occur when they are compared to field data. When an element is removed abruptly an initial, rapid increase of flow occurs that would not occur in the field. As an example, between days 303 and 304, element 189 was removed. The simulated discharge increased abruptly from 850 to 1396 gpm (3.2 to 5.3 m^3/min). In the field this block would not have been removed during one day; removal would have been a continuing process so that the change in

Table 11. Simulated Inflow and Potentiometric Surface Data at Selected Times

Time, days	Pumpage, gpm	Potentiometric elevation at selected nodes, ft			
		22	167	172	247
1	400	800	950	950	950
273	400	798	941	940	950
274	850	798	941	940	950
303	850	798	941	938	950
304	1396	798	941	938	950
405	683	797	938	925	950
555	605	796	936	909	950
660	944	796	932	890	949
860	674	794	924	844	946
945	948	794	915	822	945
1025	849	793	912	801	943
1217	755	791	894	749	939
1397	683	788	873	701	934
1483	592	787	862	679	931
1583	552	785	850	654	928
1697	563	782	836	628	924
1909	514	778	810	581	917
2121	513	773	786	537	909
2304	509	768	768	498	901
2464	511	764	752	464	894
2587	489	761	741	436	888
2709	500	758	731	404	882
2830	521	755	721	351	887
2953	495	752	711	300	871
3075	483	749	702	246	865
3207	485	746	693	182	859
3390	446	742	681	133	851
3591	455	738	669	103	842
3713	505	735	662	90.4	836
3845	526	732	654	80.2	830
3977	533	729	645	72.7	825
4119	613	726	633	66.6	819
4303	864	720	612	60.4	811
4486	683	709	557	55.9	803
4630	838	692	516	52.9	797
5003	634	632	456	47.4	783
5184	769	583	437	45.4	777
5368	695	529	422	43.6	771
5549	663	494	409	42.0	765
5733	880	349	399	40.6	759
5843	832	274	394	39.9	755

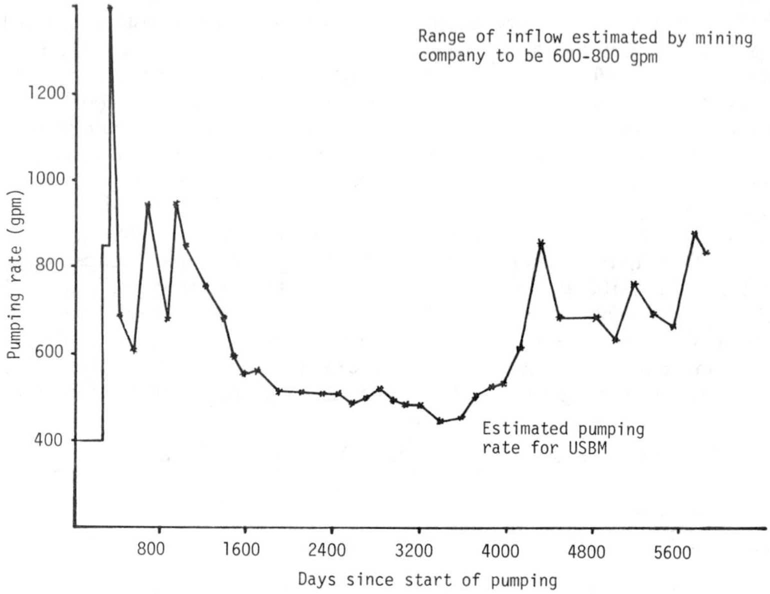

Fig. 107. Simulated pumping rate, 16-year period.

discharge would be relatively uniform. In order to minimize this problem, the data given in Table 11 were taken from the middle part of a particular computer run because this data point is more representative of actual conditions than the data at either the start or finish of a run. Fig. 108 is a plot of discharge for run 8 from 883 days to 983 days to illustrate this point. Over a long period of time this fluctuation in simulated data would average out toward the midpoint value anyway.

Another factor that should be noted is that the horizontal plane mode of UNSAT2 assumes the flow to be confined; water table conditions are not simulated. We concluded that this would not constitute a significant discrepancy from field conditions and the data appear to bear this out. Our explanation is as follows. The total thickness of the formation is about 140 ft (43 m); consequently this restriction does not become a problem until after a particular node draws down to a potential less than that at the top of the 140-ft (43 m) section. After a period of nearly four years (1431 days) only two nodes (104 and 141) were drawn down to less than 136 ft (41.5 m). The effect of this discrepancy is that the simulated discharge into the mine would perhaps be slightly larger than actual discharge because our conceptual model assumes a constant transmissivity whereas in the field the transmissivity would be reduced because of the reduced saturated thickness. The reader should remember that the mine underlies the Davis shale that is a confining layer. Therefore the natural water table does not constitute an operating variable or boundary condition.

Discussion of Simulation: The method of simulation used for this mine worked very well with few discrepancies between the simulated values and the field data. Fig. 107 shows simulated discharges. Some of the values are outside the estimated range of discharges, but this is explainable in large part by the order in which blocks were removed in simulation. The order is not necessarily the same as

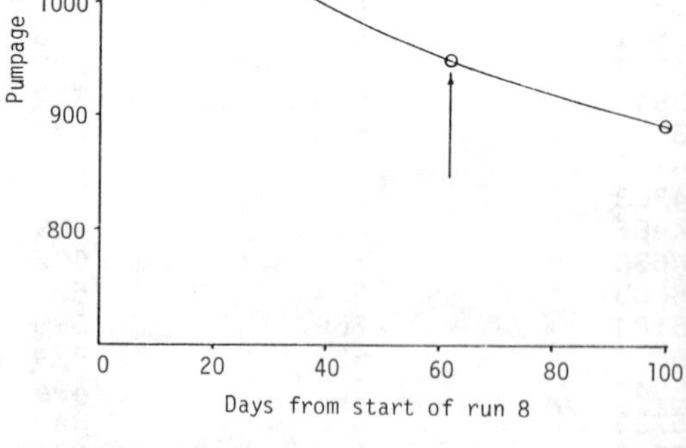

Fig. 108. Simulated pumpage during run 8. Arrow indicates value used in Table 11.

the order in which mining actually occurred; the actual order was not recorded. The fact that blocks were removed abruptly would also cause more variation in simulated flows than in field values. But the average value from simulation [660 gpm (2.5 m^3/min)] is certainly within the estimated range of field flows [600–800 gpm (2.3-3 m^3/min)].

The value of any simulation or modeling of mine inflow rests ultimately on whether the data available after initial exploration are sufficient to predict accurately the expected inflows during the life of the mine. In this mine that is certainly the case. The significant data input to the model were transmissivity and storage coefficient from rather limited well tests as well as some data on the potentiometric surface and the geometry of the ore body in terms of hydrogeologic boundaries. The inflow to the mine was simulated with enough accuracy to design a satisfactory pumping system. If the simulation had occurred prior to mining, various plans of development could have been attempted to determine the effect of each alternative on inflow and on the economics of operation. This process might have made the mining plan even more efficient with respect to mine water handling.

District Impact—Lead Mine: In this mine, as in the uranium mine, there is no hydraulic connection between the units being mined and the surface water system. This conclusion is based on the fact that there is no seasonal fluctuation of mine water inflow. In addition, there are no known water supply wells in the Bonneterre Formation. Consequently any district impact is due to the discharge of the pumped water into either surface streams or injection into other aquifers. As stated previously, if the pumped water differed in quality or quantity from the water into which it is discharged a significant effect on the present water resource could occur. This mine is in low pyrite carbonate rocks; consequently acid mine drainage is not a problem.

The benefit of using a simulation model before mining begins is the ability to estimate the quantity of water that may be discharged. Depending on the expected water quality, planning should then be done to minimize any adverse effects and to maximize beneficial effects.

8

Summary Statements

Ground water inflow into a mine and the resultant impact on district hydrology can be predicted on a site specific basis once the ground water flow systems at the site are sufficiently well understood so that a valid conceptual model can be developed. Ground water flow systems are controlled by the hydrogeologic properties of the materials at the site, including hydraulic continuity along flow paths, and boundary conditions. Preliminary delineation of hydrostratigraphic units for conceptual models frequently can be based on surface water data; specifically, spring locations and the gain-loss reaches of streams.

Quantitative predictions of mine water inflow and its impact on regional hydrology require collection and analysis of numerous types of data. Hydraulic conductivity must be defined quantitatively both parallel and perpendicular to bedding as well as areally for each hydrostratigraphic unit. The characteristics of the Davis shale in case history number 2 illustrate this concept very well. The storage characteristics also must be determined for each hydrostratigraphic unit and the potential or hydrostatic head should be quantified vertically and areally for each unit. The potentiometric measurements in the Bonneterre Formation as discussed in case history number 2 illustrate this point.

The data collection and data analysis techniques constitute the weak link in the analysis of mine inflow and district impact. The technology for characterizing a mine site in the field has lagged behind the technology for mathematical simulation; we attribute this development to the fact that computer modeling is relatively inexpensive compared to field data collection and analysis.

On-site tests provide the best means of quantifying hydraulic conductivity parallel to bedding and the storage characteristics of the unit. Pump tests with multiple observation wells provide the best information from on site tests. Hydrogeologic boundaries can be located from such multiple well tests. Multiport piezometers provide a means for determining the vertical potential in a single borehole. Borehole geophysics can facilitate the definition of hydrostratigraphic units. Vertical hydraulic conductivity is difficult but not impossible to quantify with existing field test methods; laboratory tests of cores fail to give a value representative of the field situation. The storage characteristics of a confined (artesian) hydrostratigraphic unit can be difficult to quantify when pumping changes the confined unit to an unconfined (water table) case after depressurization and dewatering.

Transmissivity and storage coefficient can be approximated to some extent for given field conditions by computer modeling if enough is known about hydrogeologic boundaries. Modeling can be used in a trial-and-error approach to match transmissivity and storage coefficients to existing water level data corresponding to estimated discharge rates. A solution can also be obtained using a technique referred to as the inverse method. Unfortunately, such a solution is seldom unique.

Fracture flow models exist but practicality usually necessitates reversion to a basic Darcy flow model. Data requirements for fracture flow models could be difficult to fulfill where fracture aperture, spacing, orientation, and length are required. Obtaining such data at depth prior to mining would be impossible for any mine simulation in fractured media. We are of the opinion that most fractured media simulations can be accomplished using conventional Darcy flow models by viewing the field problem on a macroscopic scale. The real challenge is to delineate hydrostratigraphic units properly in the field. The question of whether a broken-up fault zone is an aquifer or a barrier boundary is seldom simple. The fractures usually can be treated as an equivalent porous medium in either case.

Only a few techniques exist for predicting underground mine water inflow. These techniques frequently are designed to evaluate shaft or tunnel inflow under steady-state head conditions in a single layer rather than in multiple hydrostratigraphic units. Few computer programs exist that have the necessary characteristics needed for simulating underground mine water inflow. These necessary characteristics include seepage surface calculations, perched water, nonhomogeneous materials, anisotropy, saturated-unsaturated flow, and an irregular mesh. Programs that have these characteristics include UNSAT2, FLUMP, and TRUST. UNSAT2 was utilized herein since it is well documented and we have used it successfully in several other applications. However, as pointed out previously, it also has peculiarities and limitations.

We used the code UNSAT2 in three different ways to predict underground mine water inflow. These are:

1) Two-dimensional vertical flow to predict inflow to a unit length of drift.

2) Axisymmetric flow into a vertical shaft to determine the inflow as the shaft progresses downward.

3) Two-dimensional horizontal flow to determine the inflow to a "room-and-pillar" mode of operation in mine development. In this application we mined out elements to simulate the mining out of rooms.

The first of these applications probably is least effective because it must be assumed that the drift is extended nearly in a straight line with no curves or branching. It is also difficult to simulate any type of realistic boundary unless it is parallel to the drift. Simulation of axisymmetric flow into a shaft as the shaft proceeds downward works well except that linear boundaries are somewhat of a limiting factor. Two-dimensional horizontal flow into a "room-and-pillar" mine appears to work very well with boundaries causing no particular problem.

We have drawn the following conclusions from the results of our work on predicting mine water inflow:

1) It is not practical at the present time (late 1982) to attempt to model all aspects of flow into a mine with one large three-dimensional model. The data input—even if detailed data are available—becomes too time-consuming and eventually mesh size and storage would become a problem.

2) The only practical way to simulate inflow is to break the problem into parts and model each part.

3) It may be beneficial to use more than one of the various options of modeling at different times on the same mine. As an example, a shaft could be modeled until a region of development is reached and then a "room-and-pillar" operation could be modeled as a horizontal flow mode.

4) The most critical limiting condition for each mode of operation at least when using UNSAT2 is:

a) Axisymmetry; boundary conditions that cannot be expressed as a function of radius are difficult to handle during the simulation of axisymmetric flow into an advancing shaft.

b) Two-dimensional vertical flow; conditions that change with respect to the direction perpendicular to the plane of simulation are difficult to handle.

c) Two-dimensional horizontal flow; situations in which the vertical thickness of the aquifer is not small in comparison to the horizontal extent of the aquifers are difficult to handle.

We are uncertain about the extent to which these troublesome points are common to all models.

5) The best use of a model is to identify the most desirable method of operation when several mine plan alternatives are available. The axisymmetric evaluation of shaft inflow can be used to evaluate the effects of various radially symmetric problems or alternative mining schemes. The effect of grouting high hydraulic conductivity units can be tested along with the cost-benefit of depressurization wells. The optimal length of open shaft can be evaluated for the various hydrostratigraphic units to be penetrated. The horizontal planar mode can be used to determine inflow rates based on various drifting rates. Additional information can be obtained by altering the direction of the drifting where significant boundary conditions exist.

Additional research is needed for the purpose of defining location and limits of hydrogeologic boundaries. Techniques are needed for determining storage coefficient where multiple wells are impractical. A practical technique is needed for determining the specific yield of a confined hydrostratigraphic unit that could be dewatered during mining. Additional research is also needed on procedures capable of determining the vertical hydraulic conductivity. At present the so called *ratio method* appears to be the only practical and valid technique available and it has not been widely tried and tested. This probably is a consequence of the fact that it is expensive to perform (a minimum of three wells are required). In addition the layered case is for all practical purposes a prerequisite to its use.

Finally, we reemphasize the fact that field experience and training, as well as modeling experience and training, are required for modeling mine water inflow. The two most difficult tasks are 1) converting the hydrogeologic framework into a conceptual model that can constitute the basis for an analytical drill and test program, and 2) selecting and operating a numerical model that is compatible with the conceptual model (after revision).

9

References

Agarwal, R.G., Al-Hussainy, R., and Ramey, Jr., H.J., 1970, "An Investigation of Wellbore Storage and Skin Effect in Unsteady Liquid Flow: I. Analytical Treatment," *Society of Petroleum Engineers Journal,* Vol. 10, September, pp. 279–290.

Aguilera, R., 1976, "Analysis of Naturally Fractured Reservoirs from Conventional Well Logs," *Journal of Petroleum Technology,* Vol. 28, Pt. 2, July, pp. 764–772.

Alger, R.P., 1976, "Interpretation of Electric Logs in Fresh Water Wells in Unconsolidated Formations," Society of Professional Well Log Analysts, *Transactions,* 7th Annual Logging Symposium, Sec. CC, pp. 1–25.

American Society for Testing & Materials, 1967, *Permeability and Capillarity of Soils,* ASTM Special Technical Publication No. 417, 210 pp.

Ayatollahi, M.S., 1978, "Stress and Flow in Fractured Porous Media," Ph.D. Dissertation, Dept. of Material Science and Mineral Engineering, University of California, Berkeley, April, 154 pp.

Baltosser, R.W., and Lawrence, H.W., 1970, "Application of Well Logging Techniques in Metallic Mineral Mining," *Geophysics,* Vol. 35, pp. 143–151.

Barenblatt, G.I., Zheltov, I.P., and Kochina, I.N., 1960, "Basic Concepts in the Theory of Seepage of Homogeneous Liquids in Fissured Rocks," *Russian Journal,* PMM Vol. 24, No. 5, pp. 852–864.

Billings, M.P., 1954, *Structural Geology,* Prentice-Hall, Inc., Englewood-Cliffs, N.J., 514 pp.

Bloomsburg, G.L., 1977, "Use of Finite Element Computer Program to Determine Parallel Drain Spacing," *Proceedings,* 3rd National Drainage Symposium, American Society of Agricultural Engineers, February.

Bloomsburg, G.L., and Wells, R.D., 1978, "Seepage through Partially Saturated Shale Wastes," Final Report on Contract No. H0252065, US Dept. of Interior, Bureau of Mines.

Boulton, N.S., 1963, "Analysis of Data from Non-Equilibrium Pumping Tests Allowing for Delayed Yield from Storage," *Proceedings,* Institute Civil Engineers (London), Vol. 26, pp. 469–482.

Bouwer, H., 1966, "Rapid Field Measurement of Air Entry Value and Hydraulic Conductivity of Soil as Significant Parameters in Flow System Analysis," *Water Resources Research,* Vol. 2, pp. 729–739.

Bouwer, H., 1978, *Groundwater Hydrology,* McGraw-Hill Book Company, Inc., New York, 480 pp.

Bouwer, H., and Rice, R.C., 1976, "A Slug Test for Determining Hydraulic Conductivity of Unconfined Aquifers with Completely or Partially Penetrating Wells," *Water Resources Research,* Vol. 12, No. 3, June, pp. 423–428.

Boyle, J.M., and Saleem, Z.A., 1979, "Determination of Recharge Rates Using Temperature-Depth Profiles in Wells," *Water Resources Research,* Vol. 15, No. 6, December, pp. 1616–1622.

Brakensiek, D.L., Engleman, R.L., and Rawls, W.J., 1981, "Variation within Texture Classes of Soil Water Parameters," *Transactions,* American Society of Agricultural Engineers, Vol. 24, No. 2, pp. 335–339.

Bredehoeft, J.D., and Papadopulos, I.S., 1965, "Rates of Vertical Groundwater Movement Estimated from the Earth's Thermal Profile," *Water Resources Research,* Vol. 1, No. 2, pp. 325–328.

Bredehoeft, J.D., and Papadopulos, S.S., 1980, "A Method for Determining the Hydraulic Properties of Tight Formations," *Water Resources Research,* Vol. 16, No. 1, pp. 233–238.

Brigham, W.E., 1970, "Planning and Analysis of Pulse-Tests," *Journal of Petroleum Technology,* Vol. 22, Pt. 1, May, pp. 618–624.

Brooks, R.H., and Corey, A.T., 1964, "Hydraulic Properties of Porous Media," Hydrology Paper No. 3, Colorado State University, Fort Collins, CO.

Brown, A., 1979, "Simple Mine Inflow Evaluation for Underground Oil Shale Mines," *Mine Drainage,* G.O. Argall, Jr., and C.O. Brawner, eds., Miller-Freeman Publications, San Francisco, CA, pp. 528–555.

Brown, A., Schauer, M.I., Rowe, J.W., and Heley, W., 1977, "Water Management in Oil Shale Mining," US Bureau of Mines, RI, contract J0265019, 436 pp.

Cedergren, H.R., 1967, *Seepage, Drainage, and Flow Nets,* John Wiley and Sons, Inc., New York, 489 pp.

Chowdhury, R.N., 1974, "Generalized Steady State Field Problem and Its Solution," *Finite Element Methods in Flow Problems,* J.T. Oden, O.C. Zienkiewicz, R.H. Gallagher, and C. Taylor, eds., University of Alabama Press, Huntsville.

Chu, W.C., Garcia-Rivera, J., and Raghavan, R., 1980, "Analysis of Interference Test Data Influenced by Wellbore Storage and Skin at the Flowing Well," *Journal of Petroleum Technology,* Vol. 32, no. 1, January, pp. 171–178.

Cinco, H., Miller, F.G., and Ramey, J.H., Jr., 1975, "Unsteady-State Pressure Distribution Created by a Directionally Drilled Well," *Journal of Petroleum Technology,* Vol. 27, part 2, November, pp. 1392–1400.

Cooley, R.L., 1979, "A Method of Estimating Parameters and Assessing Reliability for Models of Steady State Groundwater Flow. 2. Application of Statistical Analysis," *Water Resources Research,* Vol. 15, No. 3, June, pp. 603–617.

Cooper, H.H., Jr., 1963, "Type Curves for Nonsteady Radial Flow in an Infinite Leaky Artesian Aquifer," *Shortcuts and Special Problems in Aquifer Tests,* R. Bentall, compiler, US Geological Survey Water Supply Paper 1545-c, pp. c48–c55.

Cooper, H.H., Jr., Bredehoeft, J.D., and Papadopulos, I.S., 1967, "Response of a Finite-Diameter Well to an Instantaneous Charge of Water," *Water Resources Research,* Vol. 3, No. 1, pp. 263–269.

Corey, A.T., 1977, "Mechanics of Heterogeneous Fluids in Porous Media," Water Resources Publication, Fort Collins, CO, 259 pp.

Crawford, G.E., Hagedon, A.R., and Pierce, A.E., 1976, "Analysis of Pressure Buildup Tests in a Naturally Fractured Reservoir," *Journal of Petroleum Technology,* Vol. 28, Pt. 2, November, pp. 1295–1300.

Croft, M.G., 1971, "A Method of Calculating Permeability from Electric Logs," US Geological Survey Professional Paper 750-B, pp. 265–269.

Dagan, G., 1978, "A Note on Packer, Slug, and Recovery Tests in Unconfined Aquifers," *Water Resources Research,* Vol. 14, No. 5, October, pp. 929–934.

Davis, J.H., 1977, "Genesis of the Southeast Missouri Lead Deposits," *Economic Geology,* Vol. 72, No. 3, May, pp. 443–450.

Davis, J.H., Rogers, R.K., and Brown, W.K., 1975 "Buick Mine (Amax Inc. & Homestake Mining Co.), Viburnum Trend, Southeast Missouri," *Guidebook to the Geology and Ore Deposits of Selected Mines in the Viburnum Trend, Missouri,* J.D. Vineyard, ed., Report of Investigations No. 58, Missouri Dept. of Natural Resources, Div. of Research and Technical Information, 56 pp.

Davis, S.N., and DeWiest, R.J.M., 1966, 1967, *Hydrogeology,* John Wiley and Sons, Inc., New York, 2nd printing May, 463 pp.

Domenico, P.A., 1972, *Concepts and Models in Groundwater Hydrology,* McGraw-Hill Book Co., Inc., New York, 405 pp.

Duguid, J.O., Jr., 1973, "Flow in Fractured Porous Media," Ph.D. Dissertation, Dept. of Civil and Geological Engineering, Princeton University, Princeton, NJ, 111 pp.

Duguid, J.O., and Abel, J.F., 1974, "Finite Element—Galerkin Method for Analysis of Flow in Fractured Porous Media," *Finite Element Methods in Flow Problems,* J.T. Oden, O.C. Zienkiewicz, R.H. Gallagher, and C. Taylor, eds., University of Alabama Press, Huntsville, pp. 599–615.

Duguid, J.O., and Lee, P.C.Y., 1977, "Flow in Fractured Porous Media," *Water Resources Research,* Vol. 13, No. 3, June, pp. 558–566.

duPrey, E.J., and Weill, L.M., 1974, "Front Displacement Model in a Fractured Reservoir," *Finite Element Methods in Flow Problems,* J.T. Oden, O.C. Zienkiewicz, R.H. Gallagher, and C. Taylor, eds., University of Alabama Press, Huntsville.

Dyck, J.H., Keys, W.S., and Meneley, W.A., 1972, "Application of Geophysical Logging to Ground-Water Studies in Southeastern Saskatchewan," *Canadian Journal of Earth Sciences,* Vol. 9, pp. 78–94.

Earlougher, R.C., 1980, "Analysis and Design Methods for Vertical Well Testing," *Journal of Petroleum Technology,* Vol. 32, No. 3, March, pp. 505–514.

Earlougher, R.C., Jr., and Kersch, K.M., 1974, "Analysis of Short-Time Transient Test Data by Type-Curve Matching," *Journal of Petroleum Technology,* Vol. 26, Pt. 2, July, pp. 793–800.

Faust, C.R., and Mercer, J.W., 1980, "Ground-Water Modeling: Recent Developments," *Ground Water,* Vol. 18, No. 6, November-December, pp. 569–577.

Franzini, J.B., 1956, "Permeameter Wall Effect," *Transactions,* American Geophysical Union, Vol. 37, No. 6, December, pp. 735–737.

Freeze, R.A., and Cherry, J.A., 1979, *Groundwater.* Prentice-Hall, Inc., Englewood-Cliffs, NJ, 604 pp.

Freeze, R.A., and Witherspoon, P.A., 1967, "Theoretical Analysis of Regional Groundwater Flow. 2. Effect of Water-Table Configuration and Subsurface Permeability Variation," *Water Resources Research,* Vol. 3, No. 2, pp. 623–634.

Frind, E.O., and Pinder, G.F., 1974, "Finite Element in the Solution of the Inverse Problem in Ground Water Flow," *Finite Element Methods in Flow Problems,* J.T. Oden, O.C. Zienkiewicz, R.H. Gallagher, and C. Taylor, eds., University of Alabama Press, Huntsville.

Frind, E.O., and Verge, M.J., 1978, "Three Dimensional Modeling of Groundwater Flow Systems," *Water Resources Research,* Vol. 14, No. 5, October, pp. 844–856.

Gale, J.E., Taylor, R.L., Witherspoon, P.A., and Ayatollahi, M.S., 1974, "Flow in Rocks with Deformable Fractures," *Finite Element Methods in Flow Problems,* J.T. Oden, O.C. Zienkiewicz, R.H. Gallagher, and C. Taylor, eds., University of Alabama Press, Huntsville.

Gillham, R.W., and Farvolden, R.N., 1974, "Sensitivity Analysis of Input Parameters in Numerical Modeling of Steady State Regional Groundwater Flow," *Water Resources Research,* Vol. 10, No. 3, pp. 529–538.

Glover, R.E., 1966, "Ground-Water Movement," Engineering Monogram 31, US Dept. of Interior, Bureau of Reclamation, 76 pp.

Goodman, R.E., Moye, D.G., van Schalkwyk, A., and Javandel, I., 1965, "Ground Water Inflows During Tunnel Driving," *Engineering Geology,* Vol. 2, pp. 39–56.

Granger, H.C., 1968, "Localization and Control of Uranium Deposits in the Southern San Juan Basin Mineral Belt, New Mexico—an Hypothesis," Professional Paper 600-B, US Geological Survey, pp. B60–B70.

Gringarten, A.C., and Ramey, H.J., Jr., 1974, "Unsteady-State Pressure Distributions Created by a Well with a Single Horizontal Fracture, Partial Penetration, or Restricted Entry," *Society of Petroleum Engineers Journal,* Vol. 14, No. 4, August, pp. 413–426.

Gringarten, A.C., Ramey, H.J., Jr., and Raghavan, R., 1974a, "Unsteady-State Pressure Distributions Created by a Well with a Single Infinite-Conductivity Vertical Fracture," *Society of Petroleum Engineers Journal,* Vol. 14, No. 4, August, pp. 347–360.

Gringarten, A.C., Ramey, Jr., H.J., and Raghavan, R., 1975, "Applied Pressure Analysis for Fractured Wells," *Journal of Petroleum Technology,* Vol. 27, Pt. 2, July, pp. 887–892.

Gringarten, A.C., and Witherspoon, P.A., 1972, "A Method of Analyzing Pump Test Data from Fractured Aquifers," *Proceedings,* Percolation through Fissured Rock, International Society for Rock Mechanics and International Association of Engineering Geology, Stuttgart, Germany, pp. T3-B1/T3-B9.

Grundmann, W.H., Jr., 1977, "Geology of the Viburnum No. 27 Mine, Viburnum Trend, Southeast Missouri," *Economic Geology,* Vol. 72, No. 3, May, pp. 349–364.

Guyod, H., 1966, "Interpretation of Electric and Gamma Ray Logs in Water Wells," *Log Analyst,* Vol. 6, No. 5, pp. 29–44.

Hackbarth, D.A., 1978, "Application of the Drill-Stem Test to Hydrogeology," *Ground Water,* Vol. 16, No. 1, January-February, pp. 5–11.

Hantush, M.S., 1959, "Nonsteady Flow to Flowing Wells in Leaky Aquifer," *Journal of Geophysical Research,* Vol. 64, No. 8, pp. 1043–1052.

Hantush, M.S., 1960, "Modification of the Theory of Leaky Aquifers," *Journal of Geophysical Research,* Vol. 65, No. 11, November, pp. 3713–3725.

Hem, J.D., 1970, "Study and Interpretation of the Chemical Characteristics of Natural Water," US Geological Survey Water-Supply Paper 1473, 2nd ed. 363 pp.

Hilpert, L.S., 1963, "Regional and Local Stratigraphy of Uranium-Bearing Rocks," *Geology and Technology of the Grants Uranium Region,* Memoir 15, Society of Economic Geologists, Uranium Field Conference, New Mexico State Bureau of Mines and Mineral Resources, 277 pp.

Ijelnfelt, A.T., Jr., and Cassidy, J.J., 1975, *Hydrology for Engineers and Planners,* Iowa State University Press, Ames, 210 pp.

Ivorslev, M.J., 1951, "Time Lag and Soil Permeability in Ground-Water Observations," Bulletin 36, Waterways Experiment Station, Corps of Engineers, US Army, April, 50 pp.

Jaeger, C., 1972, *Rock Mechanics and Engineering,* Cambridge at the University Press, New York, 417 pp.

Jargon, J.R., 1976, "Effect of Wellbore Storage and Wellbore Damage at the Active Well on Interference Test Analysis," *Journal of Petroleum Technology,* Vol. 28, Pt. 2, August, pp. 851–858.

John, E.C., and West, S.W., 1963, "Ground Water in the Grants District," *Geology and Technology of the Grants Uranium Region,* Memoir 15, Society of Economic Geologists, Uranium Field Conference, New Mexico State Bureau of Mines and Mineral Resources, 277 pp.

Johnson Division, UOP Inc., 1972, *Ground Water and Wells,* UOP Inc., St. Paul, Minnesota, 1966, 2nd printing, 440 pp.

Johnson, C.R., Greenkorn, R.A., and Woods, E.G., 1966, "Pulse-Testing: A New Method for Describing Reservoir Flow Properties between Wells," *Journal of Petroleum Technology,* Vol. 18, December, pp. 1599–1604.

Jones, P.H., and Buford, T.B., 1951, "Electric Logging Applied to Ground-Water Exploration," *Geophysics,* Vol. 16, No. 1, pp. 115–139.

Kamal, M., and Brigham, W.E., 1976, "Design and Analysis of Pulse Tests with Unequal Pulse and Shut-in Periods," *Journal of Petroleum Technology,* Vol. 28, Pt. 1, February, pp. 205–212.

Katz, D.L., et al., 1959, *Handbook of Natural Gas Engineering,* McGraw-Hill Book Co., Inc., New York, 802 pp.

Keys, W.S., and McCary, C.M., 1976, *Application of Borehole Geophysics to Water Resources Investigations of the U.S. Geological Survey,* Book 2, Chap. E-1, 126 pp.

Klute, A., 1965, *Methods of Soil Analysis,* Part 1, C.A. Black, D.D. Evans, J.L. White, L.E. Ensminger, and F.E. Clark, eds., American Society of Agronomy, Inc., Madison, WI, 770 pp.

Lambe, T.W., 1951, *Soil Testing for Engineers,* John Wiley and Sons, Inc., New York, 165 pp.

Land, L.F., 1980, "Utilizing a Digital Model to Determine the Hydraulic Properties of a Layered Aquifer," *Ground Water,* Vol. 15, No. 2, March-April, pp. 153–159.

Lankston, R.W., and Lankston, M.M., 1977, "Investigation of Geophysical and Geological Techniques to Map Thick Sedimentary Interbeds in the Columbia River Basalts," Eastern Washington University Uranium Symposium.

Linsley, R.K., Jr., Kohler, M.A., and Paulhus, J.L.H., 1958, *Hydrology for Engineers,* McGraw-Hill Book Co., Inc., New York, 2nd ed., 482 pp.

Lohman, S.W., 1972, "Ground-Water Hydraulics," US Geological Survey Professional Paper 708, 70 pp.

Lohman, S.W., et al., 1972a, "Definitions of Selected Ground-Water Terms—Revisions and Conceptual Refinements," US Geological Survey Water-Supply Paper 1988, 21 pp.

Luhr, H.P., 1974, "Organization Techniques of the Finite Element Method for Reducing Core Storage Requirements," *Finite Element Methods in Flow Problems,* J.T. Oden, O.C. Zienkiewicz, R.H. Gallagher, and C. Taylor, eds., University of Alabama Press, Huntsville.

Manula, C.B., and Owili-Eger, A.S.C., 1975, "Master Environmental Control and Mine System Design Simulator for Underground Coal Mining," US Bureau of Mines, Grant No. G0111808, 299 pp.

Matthews, C.J., and Russell, D.G., 1967, *Pressure Buildup and Flow Tests in Wells,* Monograph, Vol. 1, Society of Petroleum Engineers of AIME, New York, 172 pp.

McKinley, R.M., Vela, S., and Carlton, L.A., 1968, "A Field Application of Pulse Testing for Detailed Reservoir Description," *Journal of Petroleum Technology,* Vol. 20, March, pp. 313–321.

Melancon, P.E., 1963, "History of Exploration," *Geology and Technology of the Grants Uranium Region,* Memoir 15, Society of Economic Geologists, Uranium Field Conference, New Mexico State Bureau of Mines and Mineral Resources, 277 pp.

Millheim, K., and Cichowicz, L., 1968, "Testing and Analyzing Low Permeability Fractured Gas Wells," *Journal of Petroleum Technology,* Vol. 20, February, pp. 193–198.

Moench, R.H., and Schlee, J.S., 1967, "Geology and Uranium Deposits of the Laguna District, New Mexico," Professional Paper 519, US Geological Survey, 117 pp.

Mouat, M.M., and Clendenin, C.W., Jr., 1975, "Ozark Lead Company Mine, Viburnum Trend, Southeast Missouri," *Guidebook to the Geology and Ore Deposits of Selected Mines in the Viburnum Trend, Missouri,* J.D. Vineyard, ed., Report of Investigations No. 58, Missouri Dept. of Natural Resources, Div. of Research and Technical Information, 56 pp.

Narasimhan, T.N., and Kanehiro, B.Y., 1980, "A Note on the Meaning of Storage Coefficient," *Water Resources Research,* Vol. 16, No. 2, April, pp. 423–429.

Narasimhan, T.N., Neuman, S.P., and Witherspoon, P.A., 1978, "Finite Element Method for Subsurface Hydrology Using a Mixed Explicit-Implicit Scheme," *Water Resources Research,* Vol. 14, No. 5, October, pp. 863–877.

Narasimhan, T.N., and Palen, W.A., 1979, "A Purely Numerical Approach for Analyzing Flow to a Well Intercepting a Vertical Fracture," Society of Petroleum Engineers, California Regional Meeting, Ventura, April, pp. LBL-8816.

Narasimhan, T.N., and Witherspoon, P.A., 1977, "Numerical Model for Saturated-Unsaturated Flow in Deformable Porous Media. Part I, Theory," *Water Resources Research,* Vol. 13, No. 3, June, pp. 657–664.

Narasimhan, T.N., and Witherspoon, P.A., 1978a, "Numerical Model for Saturated-Unsaturated Flow in Deformable Porous Media. Part 3, Applications," *Water Resources Research,* Vol. 14, No.6, December, pp. 1017–1034.

Narasimhan, T.N., Witherspoon, P.A., and Edwards, A.L., 1978b, "Numerical Model for Saturated-Unsaturated Flow in Deformable Porous Media." Part 2, The Algorithm," *Water Resources Research,* Vol. 14, No. 2, April, pp. 255–261.

Neuman, S.P., 1975, "Analysis of Pumping Test Data from Anisotropic Unconfined Aquifers Considering Delayed Gravity Response," *Water Resources Research,* Vol. 11, No. 2, April, pp. 329–342.

Neuman, S.P., 1980, "A Statistical Approach to the Inverse Problem of Aquifer Hydrology. 3. Improved Solution Method and Added Perspective," *Water Resources Research,* Vol. 16, No. 2, April, pp. 331–346.

Neuman, S.P., Feddes, R.A., and Bresler, E., 1974, "Finite Element Simulation of Flow in Saturated-Unsaturated Soils Considering Water Uptake by Plants," Israel Institute of Technology, Haifi, Israel, July, 104 p.

Neuman, S.P., Fogg, G.E., and Jacobson, E.A., 1980a, "A Statistical Approach to the Inverse Problem of Aquifer Hydrology. 2. Case Study," *Water Resources Research,* Vol. 16, No. 1, February, pp. 33–58.

Neuman, S.P., and Witherspoon, P.A., 1972, "Field Determination of the Hydraulic Properties of Leaky Multiple Aquifer Systems," *Water Resources Research,* Vol. 8, No. 5, October, pp. 1284–1298.

Neuman, S.P., and Witherspoon, P.A., 1977, "Mixed Explicit-Implicit Iterative Finite Element Scheme for Diffusion Type Problems. 1. Theory," *International Journal of Numerical Methods Engineering,* Vol. 11, pp. 309–324.

Neuman, S.P., and Yakowitz, S., 1979, "A Statistical Approach to the Inverse Problem of Aquifer Hydrology. 1. Theory," *Water Resources Research,* Vol. 15, No. 4, August, pp. 845–860.

Newcomb, R.C., 1961, "Storage of Groundwater Behind Subsurface Dams in the Columbia River Basalt, Washington, Oregon, and Idaho," Professional Paper 383-A, US Geological Survey, 15 pp.

Noorishad, J., Witherspoon, P.A., and Brekke, T.L., 1971, "A Method for Coupled Stress and Flow Analysis of Fractured Rock Masses," Dept. of Geotechnical Engineering, University of California, Berkeley, Publication 71-6, March.

O'Neill, K., 1977, "Calculations of the Flow of Liquid and Heat in Fractured Porous Media," Report 77-WR-15, Water Resource Program, Princeton University, Princeton, NJ, November.

O'Neill, K., 1978, "The Transient Three-Dimensional Transport of Liquid and Heat in Fractured Porous Media," Ph.D. Dissertation, Dept. of Civil Engineering, Princeton University, January, 418 pp.

Owili-Eger, A.S.C., 1975, "Mathematical Modeling and Digital Simulation of Soil Aquifer Groundwater Motion and Mine Water Drainage," Ph.D. Dissertation, Dept. of Mining Engineering, Pennsylvania State University, 289 pp.

Paarlsberg, N.L., and Evans, L.L., 1975, "Fletcher Mine (St. Joe Minerals Corporation), Viburnum Trend, Southeast Missouri," *Guidebook to the Geology and*

Ore Deposits of Selected Mines in the Viburnum Trend, Missouri, J.D. Vineyard, ed., Report of Investigations No. 58, Missouri Dept. of Natural Resources, Div. of Research and Technical Information, 56 pp.

Papadopulos, S.S., Bredehoeft, J.D., and Cooper, H.H., Jr., 1973, "On the Analysis of 'Slug Test' Data," *Water Resources Research,* Vol. 9, No. 4, pp. 1087–1089.

Parsons, R.W., 1966, "Permeability of Idealized Fractured Rock," *Journal of the Society of Petroleum Engineers,* Vol. 6, No. 2, pp. 126–136.

Patton, F.D., 1979, "Groundwater Instrumentation for Mining Projects," *Mine Drainage,* G.O. Argall, Jr., and C.O. Brawner, eds., Miller Freeman Publications, San Francisco, CA, pp. 123–153.

Pauker, N.G., and Babushkin, V.D., 1975, "Water Flow to Mines and Foundation Pits," *Handbook for Hydrogeologists,* Vol. 1, V. M. Maximov, ed., trans. for US Bureau of Reclamation by Indian National Science Document Center, New Delhi, India, pp. 645–690.

Pirson, S.J., 1963, *Handbook of Well Log Analysis,* Prentice-Hall, Inc., Englewood-Cliffs, NJ, 326 pp.

Polubarinova-Kochina, P. Ya., 1962, *Theory of Groundwater Movement,* trans. from Russian by J.M. Roger de Wiest, Princeton University Press, Princeton, NJ, 613 p.

Prats, M., 1970, "A Method for Determining the Net Vertical Permeability near a Well from in situ Measurements," *Journal of Petroleum Technology,* Vol. 22, Pt. 1, May, pp. 637–643.

Ralston, D.R., et al., 1980, "Interactions of Mining and Water Resource Systems in the Southeastern Idaho Phosphate Field," Research Technical Completion Report Project C-7651, Idaho Water Resources Research Institute, February, 214 pp.

Ramey, H.J., Jr., 1970, "Short-Time Test Data Interpretation in the Presence of Skin Effect and Wellbore Storage," *Journal of Petroleum Technology,* Vol. 22, Pt. 1, January, pp. 97–104.

Ramey, H.J., Jr., 1975, "Interference Analysis for Anisotropic Formations—A Case History," *Journal of Petroleum Technology,* Vol. 27, Pt. 2, October, pp. 1290–1298.

Ricca, V., and Hemmerich, M., 1978, "Underground Mine Drainage Quantity and Quality Generation Model," *Proceedings,* Water in Mining and Underground Works, Asociacion Nacional de Ingenieros de Minas, Vol. 2, Granada, Spain, pp. 863–882.

Robinette, M.S., 1975, "Geophysical Investigation of Washington's Ground-Water Resources," College of Engineering Research Report 75/15-24, Washington State University, 25 pp.

Robinette, M.S., 1980, "Geophysical Techniques for Selection, Analysis and Monitoring of Waste Disposal Sites, 18th Annual Meeting, Engineering Geology and Soils Symposium, Boise, ID, April.

Robinette, M.S., 1980, "Applications of Surface and Borehole Geophysics to Uranium Waste Disposal Facilities," Preprint 80-43, SME-AIME Annual Meeting, Las Vegas, NV.

Russell, D.G., 1966, "Extensions of Pressure Build-Up Analysis Methods," *Journal of Petroleum Technology,* Vol. 18, December, pp. 1624–1636.

Sagar, B., 1978, "Galerkin Finite Element Procedure for Analyzing Flow through Porous Random Media," *Water Resources Research,* Vol. 14, No. 6, December, pp. 1035–1044.

Sahni, B.M., and Seth, H.S., 1979, "Storage Coefficients from Ground-Water Maps," *Journal of the Irrigation and Drainage Division,* American Society of Civil Engineers, Vol. 105, No. IR2, April, pp. 205–212.

Sawyer, W.K., and Mercer, J.C., 1978, "A Diffusivity Model for Fluid Flow and Heat Conduction in Po rous Media," Technical Information Center, US Dept. of Energy, Addendum, August, 150 pp.

Schlumberger Ltd., 1968, Schlumberger Technology Charts, Houston, Texas.

Schubert, J.P., 1978, "Reducing Water Leakage into Underground Coal Mines by Aquifer Dewatering," *Proceedings,* Water in Mining and Underground Works, Asociacion Nacional de Ingenieros de Minas, v. 2, Granada, Spain, pp. 911–931.

Sinclair, L.R., Fitzsimmons, D.W., and Bloomsburg, G.L., 1974, "Permeability of Unsaturated Field Soils Calculated from Laboratory Desaturation Data," *Transactions,* American Society of Agricultural Engineers, Vol. 17, No. 3.

Snow, D.T., 1968, "Rock Fracture Spacings, Openings, and Porosities," *Journal of the Soil Mechanics and Foundations Division,* American Society of Civil Engineers, Vol. 94, No. SMI, January, pp. 73–91.

Snow, D.T., 1969, "Anisotropic Permeability of Fractured Media," *Water Resources Research,* Vol. 5, No. 6, December, pp. 1273–1289.

Stallman, R.W., 1963, "Type Curves for the Solution of Single-Boundary Problems," Shortcuts and Special Problems in Aquifer Tests, compiled by R. Bentall, US Geological Survey Water Supply Paper 1545-c, pp. c45–c47.

Stallman, R.W., 1965, "Effects of Water-Table Conditions on Water Level Changes near Pumping Wells," *Water Resources Research,* Vol. 1, No. 2, pp. 295–312.

Streltsova, T.D., 1976, "Hydrodynamics of Ground Water Flow in a Fractured Formation," *Water Resources Research,* Vol. 12, No. 3, June, pp. 405–414.

Sweeney, P.H., Harrison, E., and Bradley, M., 1975, "Magmont Mine (Cominco American, Inc., and Denver Industries, Inc.), Viburnum Trend, Southeast Missouri," *Guidebook to the Geology and Ore Deposits of Selected Mines in the Viburnum Trend, Missouri;* J.D. Vineyard, ed., Report of Investigations No. 58, Missouri Dept. of Natural Resources, Div. of Research and Technical Information, 56 pp.

Su, C., and Brooks, R.H., 1976, "Hydraulic Functions of Soils from Physical Experiments and Their Applications," Report Water Resources Research Institute-41, Oregon State University.

Szilággi, G., Heinemann, Z. and Bogárd, I., 1978, "Application of a Simulation Model for a Large-Scale Karstic Water Aquifer," *Proceedings,* Water in Mining and Underground Works, Asociacion Nacional de Ingenieros de Minas, Vol. 2, Granada, Spain, pp. 951–963.

Terzaghi, K., and Peck, R.B., 1966, *Soil Mechanics in Engineering Practice,* 2nd ed., John Wiley and Sons, Inc., New York, 729 pp.

Thacker, J.L., and Anderson, K.H., 1977, "The Geologic Setting of the Southeast Missouri Lead District—Regional Geologic History, Structure and Stratigraphy," *Economic Geology,* Vol. 72, No. 3, May, pp. 339–348.

Tixier, M.P., and Alger, R.P., 1970, "Log Evaluation of Non-Metallic Mineral Deposits," *Geophysics,* Vol. 35, pp. 124–142.

Toth, J., 1963, "A Theoretical Analysis of Groundwater Flow in Small Drainage Basins," *Journal of Geophysical Research,* Vol. 68, No. 16, August, pp. 4795–4812.

US Bureau of Reclamation, 1975, *Water Measurement Manual,* 2nd ed., revised reprint, 327 pp.

US Bureau of Reclamation, 1977, "Ground Water Manual," Water Resources Technical Publication, Stock No. 024-003-00106-6, 480 pp.

Verner, E.A., Lewis, R.W., and Zienkiewicz, O.C., 1974, "Finite Elements for Two Phase Flow in Porous Media," *Finite Element Methods in Flow Problems,* J.T. Oden, O.C. Zienkiewicz, R.H. Gallagher, and C. Taylor, eds., University of Alabama Press, Huntsville.

Walton, W.C., 1962, 1967, "Selected Analytical Methods for Well and Aquifer Evaluation," Illinois State Water Survey Bulletin 49, second printing, 81 pp.

Walton, W.C., 1970, *Groundwater Resource Evaluation,* McGraw-Hill Book Co., Inc., New York, 664 pp.

Warner, D.L., Fletcher, C.S., and Cesare, J.A., 1974, "Effect of Mining Operations on Ground Water Levels in the New Lead Belt, Missouri," University of Missouri and US Dept. of Interior, Office of Water Resources Research, Project No. A-060-Mo, December, 84 pp.; available from National Technical Information Service, Springfield, VA.

Wattenbarger, R.A., and Ramey, H.J., Jr., 1969, "Well Test Interpretation of Vertically Fractured Gas Wells," *Journal of Petroleum Technology,* Vol. 21, Pt. 1, May, pp. 625–632.

Weeks, E.P., 1978, "Aquifer Tests—The State of the Art in Hydrology," *Proceedings,* International Well Testing Symposium, Lawrence Berkeley Laboratory, University of California, Berkeley, Document No. LBL-7027, 12 pp.

Wharton, H.M., 1975, "Introduction to the Southeast Missouri Lead District," *Guidebook to the Geology and Ore Deposits of Selected Mines in the Viburnum Trend, Missouri,* J.D. Vineyard, ed., Report of Investigation No. 58, Missouri Dept. of Natural Resources, Div. of Research and Technical Information, 56 pp.

White, N.F., 1968, "The Desaturation of Porous Materials," Ph.D. Dissertation, Colorado State University, Fort Collins, CO.

Williams, R.E., 1982, "Prevention of Ground Water Contamination by Uranium Mill Wastes," Geotechnical Engineering Publications, Colorado State University, Fort Collins, CO, 173 pp.

Wilson, C.R., and Witherspoon, P.A., 1974, "Steady State Flow in Rigid Networks of Fractures," *Water Resources Research,* Vol. 1, No. 2, April, pp. 328–335.

Witherspoon, P.A., Wang, J.S.Y., Iwai, K., and Gale, J.E., 1979, "Validity of Cubic Law for Fluid Flow in a Deformable Rock Fracture," Technical Information Report No. 23, Dept. of Materials Science and Mineral Engineering, University of California, and Lawrence Berkeley Laboratory, University of California, Berkeley, October, 28 pp.

Wittke, W., Rissler, P., and Semprich, S., 1972, "Three-Dimensional Laminar and Turbulent Flow through Fissured Rock According to Discontinuous and Continuous Models (in German)," *Proceedings,* Symposium on Percolation through Fissured Rock, International Society for Rock Mechanics, Stuttgart, Germany, pp. T1-H1/T1-H18.

Worley, M.T., 1962, "Groundwater Influx into a Vertical Mine Shaft," *Tansactions,* SME/AIME, Vol. 223, pp. 428–431.

Zahl, E.G., and Bloomsburg, G.L., 1980, "Seepage through Partially Saturated Soils Below a Uranium Tailings Pond," *1st International Conference on Uranium Mine Waste Disposal,* AIME, New York.

Zohdy, A.A.R., Eaton, G.P., and Mabey, D.R., 1972, "Application of Surface Geophysics to Ground-Water Investigations," Chap. D1, *Techniques of Water Resources Investigations of the U.S. Geological Survey,* Book 2, US Geological Survey, Reston, VA, 116 pp.

Appendix A

Definitions

Aquifer. A formation, group of formations, or part of a formation that contains sufficient saturated permeable material to yield significant quantities of water to wells and springs.

Artesian. Synonymous with confined; see ground water, confined.

Capillary fringe. The zone immediately above the water table in which all or some of the interstices are filled with water that is under less than atmospheric pressure and that is continuous with the water below the water table.

Conceptual model. A conceptualization of a field hydrogeologic framework that facilitates the development of analytical field testing procedures. The results of the test(s) may force changes in the initial conceptual model.

Conductivity, hydraulic. Replaces the term "field coefficient of permeability." The volume of water at the existing kinematic viscosity that will move in unit time under a unit hydraulic gradient through a unit area measured at right angles to the direction of flow, if the porous medium is isotropic and the fluid homogeneous.

Confining bed. A body of "impermeable" material stratigraphically adjacent to one or more aquifers; its hydraulic conductivity may range from nearly zero to some value distinctly lower than that of the aquifer.

Flow, steady. Any point where the magnitude and direction of the specific discharge are constant in time.

Flow, uniform. At every point the specific discharge has the same magnitude and direction.

Flow, unsteady. At any point the magnitude or direction of the specific discharge changes with time.

Fluid potential. The mechanical energy per unit mass of a fluid at any given point in space and time with respect to an arbitrary state and datum; the fluid potential is proportional to the head.

Ground water, confined. Ground water under pressure significantly greater than atmospheric; its upper limit is the bottom of a bed of distinctly lower hydraulic conductivity than that of the material in which the confined water occurs.

Ground water, perched. Unconfined ground water separated from an underlying body of ground water by an unsaturated zone.

Ground water, unconfined. Water in an aquifer that has a water table.

Head, static. Height above a standard datum of the surface of a column of water (or other liquid) that can be supported by the static pressure at a given point.

Homogeneity. Hydrogeologic properties are identical everywhere.

Hydraulic gradient. Change in static head per unit of distance in a given direction.

Hydrostratigraphic unit. A body of material having considerable lateral extent and comprising a geologic framework for a reasonably distinct hydrogeologic system.

Isotropy. Condition in which all significant properties are independent of direction.

Porosity. Property of containing interstices or voids; may be expressed quantitatively as the ratio of the volume of its interstices to its total volume.

Porosity, effective. Interconnected pore space available for fluid transmission; expressed as a percentage of the total volume occupied by the interconnecting interstices.

Potentiometric surface. A surface that represents the static head along a particular specified surface or stratum.

Specific retention. Ratio of the volume of water that the rock or soil, after being saturated, will retain against the pull of gravity to the volume of the rock or soil.

Specific yield. Ratio of the volume of water that the rock or soil, after being saturated, will yield by gravity to the volume of the rock or soil.

Storage, specific. The volume of water released from or taken into storage per unit volume of the porous medium per unit change in head.

Storage coefficient. Volume of water an aquifer releases from or takes into storage per unit surface area of the aquifer per unit change in head.

Stream, gaining. Stream or reach of a stream whose flow is being increased by inflow of ground water.

Stream, losing. Stream or reach of a stream that is losing water to the ground.

Transmissivity. Rate at which water of the prevailing kinematic viscosity is transmitted through a unit width of the aquifer under a unit hydraulic gradient.

Water table. That surface in a ground-water body at which the water pressure is atmospheric. Defined by the levels at which water stands in wells that penetrate the water body just far enough to hold standing water.

Zone, saturated. That part of the earth's crust beneath the deepest water table in which all voids, large and small, are ideally filled with water under pressure greater than atmospheric.

Zone, unsaturated. The zone between the land surface and the deepest water table; water in this zone generally is under less than atmospheric pressure, and some of the voids may contain air or other gases at atmospheric pressure.

Definitions after Lohman (1972a) except hydrostratigraphic unit and conceptual model.

Appendix B

Comparison of Methods for Inflow into Vertical Shaft*

Brown Method

A 24-ft (7.3-m) diam shaft is sunk without prior dewatering. Open at one time are 50 ft (15 m); the remainder is sealed. Advance is 5 ft per day (1.5 m/d). At "B groove" (Piceance Basin)

$$L = 50 \text{ ft}$$
$$D = \text{drawdown} = 1100 \text{ ft}$$
$$K = 0.586 \text{ ft per day}$$
$$r = 12 \text{ ft}$$
$$S_s = 3 \times 10^{-7} \text{ 1/ft}$$
$$\text{Time shaft is open} = 5 \text{ days}$$

Using equation

$$D = \frac{Q \ln R/r}{2\pi Lk} \tag{80}$$

or

$$Q = \frac{1100(2\pi)(50)(0.586)}{\ln \frac{R}{r}} \quad \text{but } \frac{R}{r}$$
$$= \sqrt{\frac{4Kt}{r_\omega{}^2 S_s}} + 1$$
$$= 32.350 \text{ cu ft per day}$$
$$= 520$$
$$168 \text{ gpm}$$

Worley Method

$$T = (0.586)(50)$$
$$= 29.30 \text{ cu ft per day}$$
$$= 219 \text{ gal per day}$$
$$S = 3 \times 10^{-7} \text{ 1/ft}$$
$$\text{Well diameter} = 6 \text{ in. assumed}$$
drawdown – 1100 ft
Shaft diameter – 24 ft
$$t = 5 \text{ days}$$

From Worley (1962), Fig. 1,

$$u = 2 \times 10^{-7}$$
$$w(u) = 6.33$$
$$Q = 150$$

Sank 25 ft in 5 days or partial penetration = 50% or $\frac{Q}{Q_0} = 0.7$

$$Q = 150\ (0.7) = 105$$

Shaft diameter factor = 1.9 This value depends on what size well was used to get data.

$$Q = 199 \text{ gpm.}$$

---

* Based on Brown (1979) and Worley (1962).

Appendix C

Relationship of Units

[Equivalent values shown in same horizontal lines. † indicates abandoned term]

A. Hydraulic conductivity

Hydraulic conductivity (K)		† Field coefficient of permeability (P_f)
Feet per day (ft day^{-1})	Meters per day (m day^{-1})	† Gallons per day per square foot (gal day^{-1} ft^{-2})
One	0.305	7.48
3.28	One	24.5
0.134	0.041	One

B. Transmissivity (T)

Square feet per day (ft^2 day^{-1})	Square meters per day (m^2 day^{-1})	† Gallons per day per foot (gal day^{-1} ft^{-1})
One	0.0929	7.48
10.76	One	80.5
0.134	0.0124	One

C. Permeability

Intrinsic permeability $k = -\dfrac{q\nu}{d\varphi/dl}$ [$(\mu m)^3 = 10^{-8}$ cm^2]	Darcy $= -\dfrac{q\mu}{dp/dl + \rho g\, dz/dl}$ [0.987×10^{-8} cm^2]	† Coefficient of permeability P or $P_m = -\dfrac{q(\text{at } 60°\text{F.})}{dh/dl}$ † [gal day^{-1} ft^{-2} at 60° F.]
One	1.01	18.4
0.987	One	18.2
0.054	0.055	One

L = length; M = mass; T = time
Lohman (1972a, Table 1, p. 18).

Appendix D

Use of an Interactive Computer System for Ground Water Modeling

INTRODUCTION

The majority of ground water flow models have been developed for large computer systems. Most of these operate under a batch mode of operation, that is, the data are input with a deck of punched cards. The advent of micro- and minicomputers over the last few years opens the possibility of adapting many of these large flow programs to the smaller computers. In general, this adaptation will require more computer time to run the program, but in many cases if the small computer is *in-house*, the overall cost may be less than that required for a large computer. There are also a number of peripheral devices that can be used to make the data input much easier. As an example, the use of a digitizer with a finite element or finite difference mesh reduces the data input time considerably; the use of a graphics terminal for display of the mesh immediately after it is developed shows immediately whether there are errors in the location of node points.

DEVELOPMENT OF AN INTERACTIVE SYSTEM

The finite element program, UNSAT2, has been used over the past several years by the authors for several ground water flow problems (Bloomsburg, 1977; Bloomsburg and Wells, 1978; Zahl and Bloomsburg, 1980). In the past these programs have been run on an IBM 370/145 computer under the batch mode of operation. One difficulty that has been encountered has been in determining the length of time step that must be specified for operation of the program. The problem is that if the time steps are too large, the program becomes unstable and operation ceases. If this happens in the middle of the problem, the only alternative is to reduce the length of time step and repeat the entire run. This problem has been alleviated in other programs by using a subroutine that determines the time step automatically. If the solution becomes unstable, the time step is reduced automatically and that portion of the solution is repeated. With an interactive system this can be done by storing the output pressures and restarting the program with the new output pressures. Under the batch mode of operation, there is an option in UNSAT2 that allows restarting of the program from the last point of solution, but this must be specified when originally starting the program.

The interactive computer system that we use currently consists of a PDP11/23 with hard disk drive, CRT terminal, plotter, graphics terminal, digitizer, and printer. The computer has 256 K bytes of core storage and the hard disk drive will accept a 5.2 M byte disk.

To run a large program on a computer such as this, some overlaying must occur; that is, subroutines are stored on disk and put into core storage only when they are actually needed for running the program. UNSAT2 required very few changes to run on this computer but some changes were necessary to take advantage of the interactive capabilities of the system. One of these changes consists of the development of a program called *DBUILD* to build the data file for running the program. This program calls for each piece of data by prompt statement and gives instructions for using the digitizer. The data are then entered in completely unformated form. The node points must be entered in order and the coordinates are determined automatically by the digitizer. The pressure readings or initial conditions on each node point are entered through the terminal. All element information also is entered through the terminal. After DBUILD is run the mesh may be displayed immediately on the graphics scope whereupon any error in node point location or the numbering of nodes for a particular element becomes immediately obvious. After the initial data file is formed, it is then possible to make any changes desired by means of the system editor.

Another program which was developed is named *FEDIT.* This program makes it possible to take the pressure readings from any complete run of UNSAT2 and put these back into the initial mesh and develop a new data file. Consequently a program can be run on UNSAT2 for a particular time and then restarted with the pressures produced by the first run. Any boundary condition can be changed; the location of node points can also be changed as long as the change does not change the number of elements or node points.

The output from the program goes into a specified data file. This file may then be reviewed on the CRT terminal and any part of it may be saved for later printing. This reduces considerably the amount of paper required for printout. Later, if it is determined that the particular data file is not needed, it may be deleted from the disk storage. Under this system if the program stops because of too large a time step, the final pressure data are filed and the program is then restarted by means of FEDIT. In this way the entire problem does not have to be rerun.

The disadvantage of using a small computer with large programs is that the running time may be quite large. Computation time on the PDP11 was approximately 4.5 min per iteration on the axisymmetric simulation of the uranium mine discussed in the foregoing text. The horizontal planar mode simulation of the lead mine required approximately 2.75 min per iteration. Generally, only one iteration per time step was required to reach a solution, although a maximum of ten iterations was allowed. The maximum number of time steps allowed was usually less than 50 per run, while over 60 runs were required for certain aspects of these simulations. However, if the computer is *inhouse,* it is entirely feasible simply to let the program run overnight. The next morning the output file may be reviewed and those portions that are necessary can then be printed out.

It is relatively easy to learn to use this type of system. A person not very familiar with computers or programming can easily learn to build, use, and edit data files in a one-week period.

Appendix E

Unsaturated Flow

The term *unsaturated flow* refers to the conditions in which soil pores are filled partially with water and partially with air. This situation constitutes a porous material filled with immiscible fluids. Immiscible fluids by definition exhibit a distinct interface when in contact with each other resulting in a pressure discontinuity across the interface due to surface tension. The fluid that tends to wet the solid particles is called the wetting fluid and the other is the nonwetting fluid. In the air-water system, air is the nonwetting fluid and water is the wetting fluid.

A somewhat analogous fluid system is that in a petroleum reservoir in which there may be up to three immiscible fluids—oil, salt water, and natural gas. The degree of wettability of the fluids is salt water → oil → gas. That is, salt water is the wetting fluid at any interface and oil is the wetting fluid at an oil-gas interface but is the nonwetting fluid at a water-oil interface. Since the oil-water and air-water systems are similar, much of the theory and methods of analyses for the air-water system or *unsaturated flow* have evolved through the petroleum literature.

In the two-fluid system the pressure difference across an interface depends on the size of the pore space in which the interface occurs. The equation commonly used to represent this dependence is

$$P_c = \frac{2\sigma}{r} \tag{81}$$

where P_c is the pressure difference, σ is the surface tension constant for the air-water interface, and r is the radius of curvature of the interface. The value of r is related directly to the size of the pore in which the interface occurs; thus the larger the pore size in which an interface occurs, the smaller the pressure difference, P_c. Another expression for P_c is

$$P_c = P_a - P_w \tag{82}$$

where P_c is the capillary pressure, P_a is the air pressure (often assumed to be atmospheric), and P_w is the pressure in the water. Capillary pressure is also called tension or negative pressure. In a static water table situation the capillary pressure head $P_c/\rho g$ is also the vertical distance above the water table.

Because P_c is related to the pore size in which interfaces occur, it is also related to whether a particular size pore is filled with water or air. This relationship is also expressed by the curve shown in Fig. 109 where P_d is the displacement pressure.

The horizontal axis of this plot represents a zero capillary pressure head or the water table, so the curve actually gives the saturation distribution above the water table under static conditions. The value of capillary pressure head equal to $P_d/\rho g$ is the displacement pressure head and is an index of the larger pore sizes in the porous media. The residual saturation, S_r, is an experimental coefficient

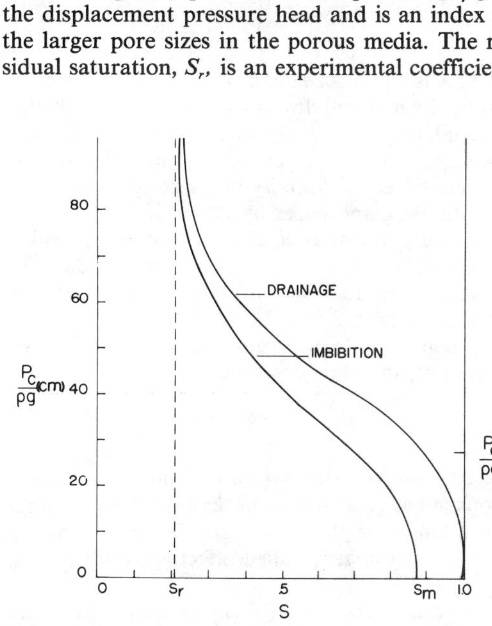

Fig. 109. Typical capillary pressure-saturation curve.

that represents the amount of water the soil will hold against the pull of gravity, while S_m is the maximum saturation during imbibition.

The drainage curve is obtained when the porous medium is saturated initially and the capillary pressure is then increased in stages and the material desaturates. The imbibition curve represents the reverse process, with the material initially dry and then imbibing water. The difference between the two curves is caused largely by trapped air. This air will eventually (up to several months) dissolve in the water and the material will become saturated. The importance of this in field problems is that the soil is seldom completely saturated, particularly in infiltration situations.

The effective saturation, S_e, is defined by

$$S_e = \frac{S - S_r}{1 - S_r} \tag{83}$$

for drainage or

$$S_e = \frac{S - S_r}{S_m - S_r} \tag{84}$$

for imbibition.

An empirical relationship between S_e and P_c that was introduced by Brooks and Corey (1964) is

$$S_e = \left(\frac{P_d}{P_c}\right)^{\lambda} \tag{85}$$

where λ is a pore size distribution index. The more readily the material drains with respect to a change of capillary pressure, the larger is λ and the narrower is the range of pore sizes. The value of λ is determined experimentally by plotting S_e vs. P_c on logarithmic graph paper as shown in Fig. 110.

The experimental coefficient $P_d/\rho g$, S_r, and λ are the only soil parameters required to describe capillary pressure—saturation-relative conductivity relationships.

When considering water flow at unsaturated conditions, the Darcy equation,

$$q = -K\Delta H/\Delta L$$

is used; however, the hydraulic conductivity is not a constant as at saturation but is a function of either saturation or capillary pressure. Since K is a variable, it is ordinarily called effective conductivity, K_e.

A relative conductivity can now be defined as

$$K_r = \frac{K_e}{K} \tag{86}$$

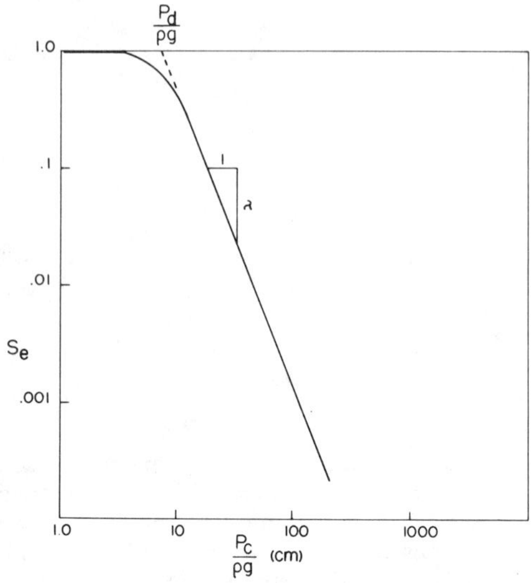

Fig. 110. Typical capillary pressure-effective saturation curve.

where K is the saturated conductivity. The value of K_e may now be expressed as $K_e = K_r K$, with K_r a function of capillary pressure or saturation that varies from 0 to 1.0. A relationship between K_r and capillary pressure used by Brooks and Corey (1964) is

$$K_r = \left(\frac{P_d}{P_c}\right)^{2 + 3\lambda} \tag{87}$$

A plot of this relationship on logarithmic paper (Fig. 111) appears very similar to that for effective saturation.

It should be pointed out that the equations given for S_e and K_r describe only the straight line portion of the curve when $P_c > P_d$. For $P_c < P_d$, $K_r = 1.0$ and $S_e = 1.0$.

These two straight line relationships ignore the curved transition portion of the curve near P_d. This is a weakness with using these relationships, but there is some evidence that the curved transition is a laboratory phenomenon and under a dynamic situation in the field, data would very nearly fit the two straight lines.

In general, the given relationships work very well for granular media such as sedimentary materials. They do not work nearly so well for agricultural soils that have *structure* or worm holes and root holes. Since these relationships are reasonably accurate for sedimentary materials, the discussion of measurement of properties of an unsaturated zone is pointed toward obtaining the parameters $P_d/\rho g$, λ, S_r, and the saturated conductivity, K. In

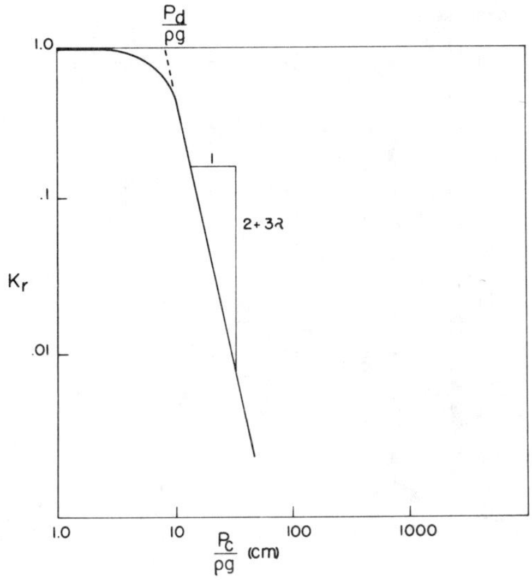

Fig. 111. Capillary pressure, relative conductivity curve.

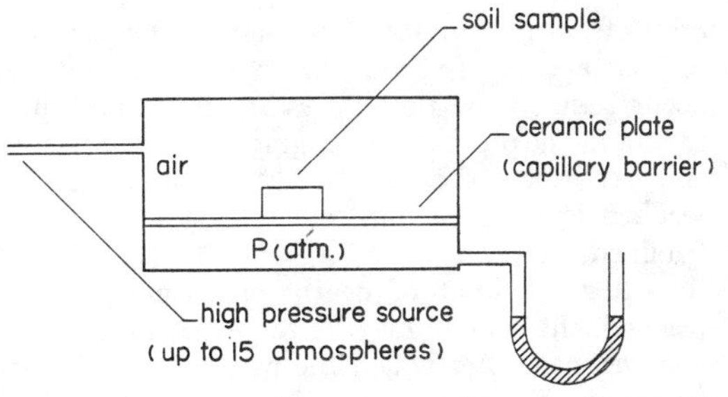

Fig. 112. Schematic diagram of pressure plate apparatus.

addition, it may sometimes be necessary to determine the porosity. Since measurement of porosity is a fairly standard soil mechanics laboratory process, it will not be discussed here.

CAPILLARY PRESSURE—SATURATION MEASUREMENTS

The methods discussed herein are laboratory methods because there are no completely satisfactory field methods for obtaining these data. Basically all methods involve imposing a negative pressure in the water with respect to the air and then determining the water saturation at a particular capillary pressure.

Pressure Plate

A schematic diagram of the pressure plate apparatus is shown in Fig. 112. This device is used by soil physicists to determine the capillary pressure-saturation relationship for agricultural soils. The soil sample may be encased in plastic so that it may be handled and still the bottom will be in contact with the capillary barrier. The barrier must be saturated so that the high pressure above the barrier is not transmitted to the chamber below the barrier. The high pressure forces water out of the pores in the soil but not out of the smaller pores in the barrier. Note that the capillary pressure or tension is equal to the difference between the air pressure and water pressure. Therefore, capillary pressure can be induced by either increasing the air pressure or decreasing the liquid pressure.

The procedure for obtaining data is to saturate the sample, place it in the pressure plate, and set the desired pressure. After enough water has flowed from the sample for the saturation to equilibrate with the pressure that was set, the sample is removed and weighed so that the saturation value may be determined. Then the sample is replaced, another pressure is set, and the process is continued to get as much data as desired. The pressure plate has the advantage of allowing tensions or capillary pressures greater than one atmosphere that other methods do not allow.

Capillary Pressure Cell

A schematic diagram of a capillary pressure cell is shown in Fig. 113. There have been many forms of this device, but this is the simplest and most accurate and rapid to use.

The method of operation of this equipment is to place the soil sample in the plastic cup and vacuum saturate it. It is then placed on the balance and connected to the horizontal glass tubing with flexible plastic tubing. The weight is then recorded, being careful that there is no free standing water on the surface of the sample; next the capillary pressure is set and the water-air interface will move along the glass tube. When the interface stops moving equilibrium has been reached and the weight is again recorded. After the interface stops moving, it will soon begin moving backward as water evap-

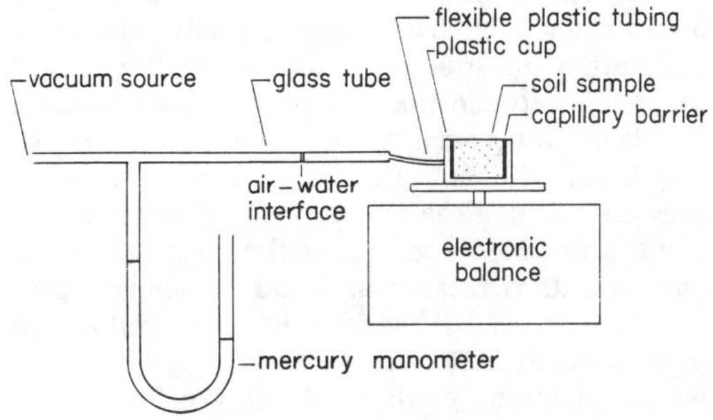

Fig. 113. Diagram of capillary pressure cell.

orates from the surface of the sample. Leaving the soil surface exposed and allowing evaporation to occur reduces the time necessary for equilibrium to occur. The process of changing capillary pressure and recording the weight after equilibrium is reached is repeated until enough data points are obtained.

After the limit of nearly one atmosphere is reached, the sample may be removed, oven dried, and weighed, after which the porosity may be determined. The various weights are then used to calculate the saturation.

This method is usually used to define the drainage curve but Su and Brooks (1976) used similar equipment to determine the imbibition curve. In their procedure water was added to the soil sample to bring about a lower capillary pressure.

As previously mentioned there are variations on these techniques that have been used in the past, but the methods described here appear to be the best at this time. In either of these methods *undisturbed* samples removed with a coring device can also be used if the device is encased in plastic tubing.

Capillary Pressure-Relative Conductivity Measurements

The most common steady state method for determining capillary pressure-relative conductivity relationships is called the *short column* method. A diagram of this device is shown in Fig. 114. The soil sample (if unconsolidated) is packed into a plastic column several inches in diameter and 12 to 15 in. (30.5 to 38 cm) in length. There are at least two tensiometers along the cylinder to measure the capillary pressure and pressure gradient and there are capillary barriers at both ends of the column. A semi-consolidated material in the form of a core may be encased in heat shrinkable tubing with tensiometers and capillary barriers clamped onto the core.

The procedure used initially is to vacuum saturate the entire column. It is then mounted on a rack to hold it in a vertical position and the tensiometer and inflow and outflow tubes are connected. The outflow and inflow tubes are set to give a very slight capillary pressure initially. The inflow and outflow must be set so that the hydraulic head difference between the tensiometers is the same as the distance between the tensiometers. This gives a hydraulic gradient of 1.0 and a constant capillary pressure through the sample. After the proper gradient is obtained and steady flow is reached, the outflow rate is measured. Another capillary pressure is then set by lowering both the inflow and outflow tubes. After steady flow is again reached with a hydraulic gradient of 1.0, the flow rate is remeasured. This process is repeated until the max-

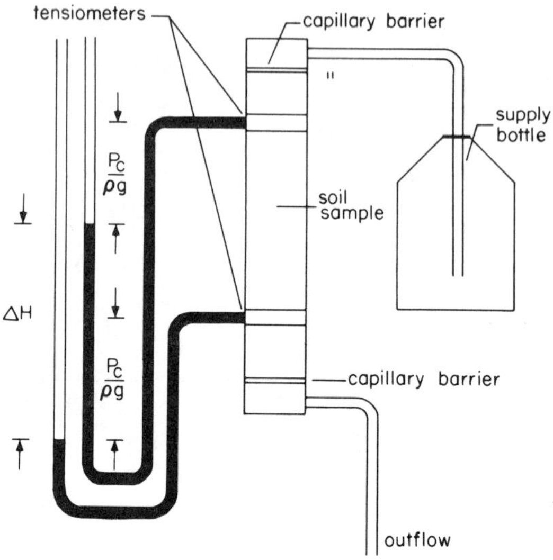

Fig. 114. Capillary pressure-relative conductivity apparatus.

imum desired capillary pressure is reached. After the sample begins to desaturate, the outflow rate drops considerably. The saturated conductivity value is calculated from the first setting by using the Darcy equation, and the effective conductivity values are calculated for each succeeding value of capillary pressure. Relative conductivity is then calculated and the data for the curve shown in Fig. 111 is obtained.

With some materials (particularly fine grained) the foregoing procedure is very time-consuming and frustrating, as days are required for equilibrium to be attained. For this reason several methods have been developed with which to calculate relative conductivity values from capillary pressure-saturation data. The simplest of these is from the previously mentioned Brooks-Corey relationships that are satisfactory for granular material. A method developed by Sinclair (1974) works reasonably well for materials with structure, such as agricultural soils.

In general, for most calculations it is satisfactory to calculate the relative conductivity but not the saturated conductivity.

An unsteady state procedure was developed by Corey (1977) to measure the parameters: porosity, maximum conductivity, residual saturation, displacement pressure, and pore size distribution index. The sample used for this procedure was similar to the *short column* previously described except that the inflow rate varied with time and a gamma ray source and detector were mounted along the column to measure the saturation. When the data from this procedure are plotted on the capillary pressure-conductivity curve, there is no transition region

Table 12. Brooks and Corey Parameters Estimated from Soil Moisture-Capillary Pressure Data (after Brakensiek, et al., 1981, Tables 3 and 4)

Texture	N sample size	λ (−)	$\ln \frac{P_d}{\rho g}$, cm	ϕ (−)	θr (−)	$\ln \frac{K}{2}$, cm/s
Sand	19	0.739	2.853	0.349	0.017	−4.780
Loamy sand	69	0.670	2.273	0.410	0.024	−3.818
Sandy loam	166	0.615	2.820	0.423	0.048	−5.248
Loam	83	0.496	3.144	0.452	0.034	−6.314
Silt loam	199	0.455	3.769	0.484	0.018	−7.693
Sand clay loam	129	0.587	3.253	0.406	0.075	−6.592
Clay loam	112	0.509	3.305	0.476	0.087	−6.844
Silty clay loam	175	0.405	3.607	0.473	0.054	−8.067
Silty loam	26	0.431	3.302	0.476	0.085	−7.331
Clay	108	0.432	3.494	0.475	0.106	−8.018

near the displacement pressure as was discussed previously. Since this is a dynamic process, the situation in the field, many people feel the transition region is largely a laboratory phenomenon resulting from the steady state procedure.

At the present time there are no satisfactory field methods for determining all the soil parameters. Bouwer (1966) developed a field procedure for measuring displacement pressure and saturated conductivity, but the pore size distribution index is also needed. In the future it may be possible to make use of the many numerical solutions to unsteady flow problems to develop such procedures.

Equations have been developed recently for estimating the hydraulic properties of porous media from the percent sand, silt, and clay in soil (American Society for Testing & Materials, 1967). Brakensiek (1981) correlated the hydraulic properties of the several textural classes of soils with the percent sand and clay and obtained the values shown in Table 12. All correlation coefficients were greater than 0.828, showing reasonably good correlation.

In Table 12 ϕ is porosity and the other symbols are as defined previously. These values were developed from the analysis of 1085 soil samples of 10 soil textures from throughout the country. They were all unconsolidated soils. The hydraulic properties of sandstone materials (consolidated) were estimated in the present project based on data obtained from the literature (White, 1968). Each material in the project was matched on the basis of porosity and saturated conductivity with a material in that reference.

In most ground water flow problems it is necessary to obtain unsaturated flow values as described previously if no better data are available. Saturated hydraulic conductivity and storage coefficient are usually available if on-site pump tests have been conducted of each hydrostratigraphic unit.

In any situation in which the possibility of the material desaturating exists, it is preferable to use even approximate values of the hydraulic properties rather than to use a saturated flow model.

Appendix F

Simulation Summaries of the Uranium and Lead Mines

Simulation Output, Uranium Mine

Run No.	Total time, days	Piezometeric head node 194, ft	Depressurization flow, gpm	Shaft inflow, gpm	Drift inflow, gpm		Total Q
					Upper	Lower	
0	0	1966	—				
1	20	1894	800				
2	26	1823	1100				
3	30	1722	1700				
4	34	1593	2250				
5	38	1440	2800				
6	46	1131	3400				
7	51	951	3750				
8	69	640	3400				
9	74	609	3100				
10	89	558	2770				
11	104	539	2500				
12	119	532	2300				
13	135	524	1800	327	—	—	327
14	150	564	1500	364	—	—	364
15	166	541	1500	407	—	—	407
16	181	535	1380	452	—	—	452
17	186	533	1300	534	—	—	534
18	187.5	528	1300	555	—	—	555
19	189.5	524	1300	566	—	—	566
20	191.5	526	1300	519	—	—	519
21	193.5	528	1450	382	—	—	382
22	195.5	531	1450	327	—	—	327
23	197.5	525	1600	281	—	—	281
24	199.5	517	1600	312	—	—	312
25	202	513	1500	352	—	—	352
26	204	524	1450	261	—	—	261
27	205.5	517	1450	364	—	—	364
28	225.5	506	1250	262	610	—	872
29	240.5	460	1425	234	803	—	1037
30	242.5	458	1350	173	922	—	1095
31	244.6	448	1350	219	963	—	1182
32	246.0	445	1350	191	979	—	1170
33	247.8	440	1350	201	1008	—	1209
34	249.8	434	1350	214	1017	—	1231
35	251.8	428	1350	221	1023	—	1244
36	254.3	423	1325	231	1056	—	1287
37	256.3	425	1325	171	1097	—	—

Simulation Output, Uranium Mine (cont'd)

Run No.	Total time, days	Piezometeric head node 194, ft	Depres-surization flow, gpm	Shaft inflow, gpm	Drift inflow, gpm Upper	Drift inflow, gpm Lower	Total Q
38	257.8	419	1325	236	1093	—	1329
39	259.3	410	1325	158	1086	258	1502
40	261.3	410	1190	121	1113	385	1619
41	262.8	409	1190	96.8	1132	401	1630
42	265.8	401	1325	—	1157	480	1637
43	270.8	391	1325	—	1201	560	1761
44	275.8	368	1475	—	1244	620	1864
45	280.8	366	1272	—	1310	700	2010
46	285.8	362	1375	—	1374	793	2167
47	290.8	357	1375	—	1438	884	2322
48	300.8	349	1375	—	1522	1048	2570
49	310.8	341	1375	—	1599	1164	2763
50	320.8	342	1325	—	1623	1294	2917
51	330.8	338	1325				
52	340.8	333	1325				
53	350.8	333	1300				
54	360.8	330	1300				
55	370.8	326	1300				
56	380.8	327	1275				
57	390.8	329	1250				
58	400.8	328	1250				
59	410.8	318	1300				
60	420.8	310	1325				
61	425.8	308	1325				
62	438.8	400	1045				
63	448.8	389	955				
64	470.1	589					
	498.8	642					
65	520.1	661					
	548.8	679					
66	570.1	689					
	598.8	699					

Potentiometric Surface of Upper and Lower Drifts, Uranium Mine

		Node No.														
		24	48	96	120	159	187	201	222	281	289	293	297	301	305	309
		Distance, ft														
Time, days	Q, gpm	0	10	30	50	130	250	450	750	1100	300	5000	8000	11000	14000	18000
								Upper Drift								
0.1	8.79	797	798	808	825	931	1107	1357	1643	1877	2171	2206	2210	2210	2210	2210
3.2	1.12	206.2	206.4	207.7	211	230	257	300	372	464	939	1358	1802	2044	2156	2210
5.65	1.08	204.0	204.2	206.3	209	221	243	280	337	403	757	1098	1524	1831	2036	2210
8.96	1.00	197.3	197.6	199.8	202.7	214	231	260	306	359	645	931	1317	1643	1911	2210
12.27	0.916	190.5	190.8	192.4	194.6	207.5	223	249	289	336	589	346	1209	1541	1841	2210
19.25	0.809	179.0	179.4	182.1	185.0	198.8	217	240	275	317	—	—	—	—	—	—
20.58	0.781	174.9	176.1	178.8	182.7	197.9	215	238	274	315	538	770	1112	1447	1776	2210
26.19	0.730	171.0	171.5	174.1	178.0	193.5	210	233	268	308	527	755	1094	1431	1765	2210
56.97	0.574	144.6	148.7	160.5	166.5	183.0	201	225	259	299	516	742	1081	1419	1757	2210
68.23	0.540	142.6	146.6	158.3	164.0	180.7	200	224	258	297	514	740	1079	1417	1756	2210
107.84	0.518	140.9	144.7	156.3	163.0	180.5	200	224	258	297	514	740	1079	1417	1756	2210
111.27	0.518 steady state 23.02															
								Lower Drift								
0.1	6.97	1174	1175	1178	1185	1237	1353	1543	1766	1949	2179	2207	2210	2210	2210	2210
3.0	1.28	239.2	239.3	240.1	241.6	253.9	282.7	336.5	418.3	513.0	1002	1427	1855	2074	2168	2210
5.54	1.10	212.9	212.9	213.3	214.2	221.3	240.3	276.3	332.8	398.9	753	1095	1520	1828	2033	2210
12.76	0.978	194.6	194.7	195.3	197.1	207.3	219.4	244.2	283.8	329.8	578.6	833	1194	1526	1831	2210
20.04	0.917	182.7	182.8	183.5	185.0	196.1	210.2	233.6	268.2	309.6	534.5	768	1111	1447	1776	2210
29.33	0.857	168.3	168.5	170.3	172.5	184.6	204.2	228.0	262	—	—	—	—	—	—	—
34.07	0.827	162.4	162.5	163.5	165.3	179.4	199.4	221.9	256.2	296.3	514.7	743	1083	1421	1758	2210
43.35	0.767	140.5	140.8	145.5	149.9	172.3	193.6	219.6	253.3	293.2	510.4	738	1077	1417	1756	2210
53.51	0.723	128.3	128.6	131.4	138.1	166.9	191.3	218.8	252.5	292.3	509.4	737	1076	1416	1755	2210
66.24	0.675	117.4	117.8	121.0	128.1	159.4	182.4	209.3	244.0	284	505	734	1074	1415	1754	2210
98.09	0.587	96.4	96.9	101.2	109.6	143.9	177.3	207.7	242.3	282.3	500.3	729	1070	1411	1752	2210
100.37	0.585 steady state 11.893	96.0	96.6	100.6	109.3	144.2	177.1	207.7	242.3	282.3	500.3	729	1070	1413	1753	2210

Simulation Output, Lead Mine

				Node				
	Δt	Σt	Pumpage	22	104	167	172	247
1	1	1.0	400	800	914	950	950	950
		2.3	400	800	917	947	949	950
		100.0	400	799.8	905	945.6	947.6	950
2	1	101.0	400	800.0	904.7	946	948.0	950
	52	152.0	400	799	888.7	949.1	946.1	950
	102	202.0	400	798.5	871.9	942.7	944.1	950
	173	273.0	400	797.8	848.3	941.0	940.3	949.9
3	1	274.0	850	798.0	847.7	940.1	939.9	950.0
	10	283.0	850	797.9	844.9	940.8	939.4	950.0
3A	1	284.0	850	798.0	844.7	941.0	939.0	950.0
	9.0	292.0	850	797.9	842.1	940.8	938.5	950.0
	20.0	303.0	850	797.8	838.2	940.6	937.9	950.0
4	1	304.0	1396	798.0	837.6	941.0	937.9	950.0
	52	355.0	868	797.6	766.8	939.9	934.4	949.9
	102	405.0	747	797.2	708.2	938.9	929.9	949.8
	150	453	683	796.9	668.3	937.9	924.5	949.6
5	1	454	678	796.9	667.3	937.9	923.9	950.0
	52	505	638	796.6	635.0	936.9	917.0	949.7
	102	555	605	796.0	609.0	935.7	909.1	950.0
	155	608	575	796.0	585.8	934.3	899.6	949.0
6	1	609	1494	796.0	584.5	934.0	899.8	949.0
	52	660	944	795.7	409.4	932.4	889.6	948.5
	100	708	824	795.4	352.8	930.7	879.6	948.0
7	1	709	817	795.0	352.0	930.9	879.9	948.0
	52	760	756	795.0	320.9	928.7	868.4	947.4
	102	810	710	794.7	300.3	926.3	856.6	946.7
	152	860	674	794.4	284.8	923.6	844.2	945.8
	175	883	658	794.2	278.8	922.3	838.4	945.6
8	1	884	1210	794.0	276.0	921.9	837.8	946.0
	32	915	1021	793.9	203.5	920.0	829.9	945.4
	62	945	948	793.7	181.1	918.0	822.1	944.9
	100	983	892	793.3	167.9	915.3	812.1	944.1
9	1	984	886	793.0	167.7	914.9	811.7	944.0
	42	1025	849	793.0	158.7	911.9	800.8	943.1
	82	1065	821	792.7	152.2	908.7	790.0	942.3
10	1	1066	887	792.9	151.9	908.9	789.7	942.0
	32	1097	837	792.5	147.7	906.1	781.4	941.3
	62	1127	810	792.1	143.9	903.3	773.2	940.6
	92	1157	794	791.7	140.4	900.3	765.1	939.9
	122	1187	771	791.3	137.3	897.2	757.0	939.2
	152	1217	755	790.9	134.6	894.0	748.8	938.5
	180	1245	741	790.5	132.2	891.0	741.3	939.0
11	1	1246	738	791.0	131.9	890.9	740.7	938.0
	52	1297	719	789.9	128.2	885.0	727.0	936.6
	102	1347	700	789.1	124.8	879.0	713.8	935.2
	152	1397	683	788.2	121.8	873.0	700.7	933.8
	186	1431	672	787.6	119.9	868.8	691.9	932.8
12	1	1432	679	788.0	117.1	868.9	691.7	933.0
	52	1483	592	786.8	75.8	862.4	678.8	931.4
	100	1531	569	785.8	72.5	856.4	666.9	929.9
13	1	1532	567	786.0	72.5	855.9	666.8	930.0
	52	1583	552	784.7	70.4	849.7	654.4	928.3
	114	1645	536	783.4	68.5	842.0	639.6	926.3

Simulation Output, Lead Mine (cont'd)

				Node				
	Δt	Σt	Pumpage	22	104	167	172	247
	1	1646	606	783.0	68.5	841.9	639.8	926.0
14	52	1697	563	782.2	67.1	835.7	627.7	924.3
	90	1735	—	—	—	—	—	—
	1	1736	631	781.0	66.2	830.9	618.8	923.0
15	52	1787	—	780.2	65.0	824.7	607.2	921.0
	122	1857	—	—	—	—	—	—
	1	1858	526	779.0	63.6	815.9	591.8	919.0
16	52	1909	514	777.8	62.6	810.2	580.9	917.1
	122	1979	499	776.1	61.4	802.4	566.4	914.5
	1	1980	757	776.0	61.4	801.9	565.8	914.0
17	52	2031	561	775.0	60.6	796.3	555.6	912.0
	90	2069	542	774.0	60.0	792.2	547.9	910.6
	1	2070	526	774.0	60.0	791.9	547.8	911.0
18	52	2121	513	772.7	59.1	786.5	537.0	908.9
	91	2160	504	771.8	58.6	782.4	528.7	907.4
	1	2161	626	772.0	58.6	781.9	528.8	907.0
19	52	2212	552	770.6	56.4	776.8	517.9	904.9
	92	2252	529	769.6	55.4	772.7	509.4	903.3
20	52	2304	509	768.5	54.5	767.8	498.3	900.9
	92	2344	497	767.5	53.9	763.8	490.0	899.2
21	1	2345	510	767.0	53.9	763.9	489.8	899.0
	31	2375	498	766.5	53.4	760.9	483.6	897.7
22	1	2376	508	767.0	53.4	760.9	483.8	898.0
	28	2403	498	766.1	53.0	758.3	478.1	896.8
23	1	2404	694	766.0	53.0	757.9	477.8	897.0
	61	2464	511	764.3	52.3	752.3	464.9	898.0
24	1	2465	585	764.0	52.3	751.9	464.8	894.0
	61	2525	509	762.7	51.4	746.5	450.7	891.4
25	1	2526	518	763.0	51.4	746.9	450.8	891.0
	62	2587	489	761.3	50.6	741.3	436.0	888.3
26	1	2588	487	761.0	50.6	740.9	435.8	888.0
	61	2648	469	759.5	49.9	734.9	421.2	885.4
27	1	2649	642	760.0	49.9	735.9	420.8	885.0
	61	2709	500	758.1	49.3	730.9	404.0	882.4
28	1	2710	513	758.0	49.3	730.9	403.7	882.0
	31	2780	494	757.4	49.0	728.4	394.0	880.7
29	1	2741	696	757.0	49.0	727.9	393.7	880.9
	29	2769	575	756.5	48.8	725.7	381.8	879.7
30	1	2770	566	756.0	48.8	725.9	381.5	880.0
	61	2830	521	754.9	48.2	721.1	350.5	877.3
31	1	2831	518	755.0	48.2	720.9	350.5	877.0
	61	2891	497	753.7	47.7	716.3	323.0	874.2
32	1	2892	543	754.0	47.7	713.9	322.7	874.0
	62	2953	495	752.2	47.2	711.4	299.7	871.3
33	1	2954	570	752.0	47.2	710.9	299.6	871.0
	61	3014	492	750.7	45.7	706.6	277.8	868.4
34	1	3015	563	750.9	45.7	706.9	277.6	868.0
	61	3075	483	749.4	45.2	702.4	246.0	865.4
35	1	3076	497	749.0	45.2	701.9	245.3	865.0
	31	3106	432	748.4	44.9	699.9	226.8	863.7

Simulation Output, Lead Mine (cont'd)

	Δt	Σt	Pumpage	Node 22	104	167	172	247
36	1	3107	525	748.0	44.9	699.9	226.5	863.9
	59	3165	503	747.0	44.5	696.0	198.4	861.4
37	1	3166	502	747.0	44.4	695.9	197.6	861.0
	42	3207	485	746.1	43.5	693.2	181.8	859.2
	91	3256	476	745.1	43.2	690.0	165.9	857.1
38	1	3257	480	745.0	43.2	689.9	165.7	857.0
	42	3298	467	744.0	42.9	687.3	154.3	855.2
	92	3348	453	743.0	42.5	684.0	142.3	853.0
39	1	3359	448	743.0	42.5	683.9	141.8	853.0
	42	3390	446	742.1	42.2	681.3	133.3	851.2
	92	3440	434	741.0	41.9	678.2	124.2	849.0
40	1	3441	454	741.0	0	677.9	123.8	849.0
	31	3471	437	740.3	0	676.1	119.0	847.7
41	1	3472	430	740.0	0	675.9	118.8	848.0
	59	3530	419	738.8	0	672.5	110.4	845.4
42	1	3531	503	739.0	0	672.0	109.9	845.0
	61	3591	455	737.6	0	668.8	102.6	842.3
43	1	3592	557	738.0	0	668.9	102.9	842.0
	61	3652	497	736.5	0	665.4	96.2	839.4
44	1	3653	561	736.5	0	665.0	96.1	839.0
	61	3713	505	735.1	0	661.7	90.4	836.4
45	1	3714	502	735.0	0	661.9	90.3	836.0
	42	3755	487	734.2	0	660.9	86.9	834.3
	92	3805	471	733.2	0	656.6	83.1	832.2
46	1	3806	604	733.0	0	656.9	83.0	832.0
	42	3845	526	732.3	0	654.3	80.2	830.2
	90	3895	500	731.3	0	651.2	77.2	828.2
47	1	3896	591	731.0	0	650.9	77.1	828.0
	82	3977	533	729.5	0	645.5	72.7	824.6
	122	4017	517	728.6	0	642.4	70.8	822.9
48	1	4018	695	729.0	0	641.9	70.8	823.0
	102	4119	613	726.4	0	632.9	66.6	818.8
	153	4170	587	725.0	0	627.5	64.7	816.6
49	1	4171	1491	725.0	0	627.9	64.7	817.0
	42	4211	989	723.7	0	623.2	63.3	815.2
	91	4261	769	722.0	0	617.6	61.7	813.2
50	1	4262	—	722.0	0	617.9	61.7	813.0
	42	4303	864	720.3	0	611.9	60.4	811.3
	91	4352	776	717.7	0	600.1	59.1	809.3
51	1	4453	780	717.9	0	599.7	59.1	809.0
	42	4494	742	715.3	0	587.2	58.0	807.3
	92	4444	703	711.9	0	570.8	56.8	805.3
52	1	4445	705	711.9	0	570.7	56.8	805.0
	42	4486	683	708.9	0	557.4	55.9	803.3
	92	4536	658	705.2	0	542.3	54.8	801.4
53	1	4537	2191	704.9	0	541.7	54.8	801.0
	102	4638	838	692.4	0	515.8	52.9	797.1
	181	4717	730	677.8	0	499.1	51.5	794.0
54	1	4718	721	677.8	0	498.8	51.5	794.0
	102	4819	682	659.8	0	480.9	49.9	790.2
	184	4901	656	646.9	0	468.8	48.7	787.1

Simulation Output, Lead Mine (cont'd)

				Node				
	Δt	Σt	Pumpage	22	104	167	172	247
	1	4902	661	646.8	0	468.9	48.7	787.0
55	102	5003	634	632.5	0	455.9	47.4	783.2
	181	5082	616	622.5	0	447.1	46.5	780.4
	1	5083	1727	621.9	0	446.9	46.5	780.0
56	102	5184	769	583.0	0	437.0	45.4	776.5
	184	5266	721	554.7	0	429.9	44.5	773.7
	1	5267	728	554.7	0	429.9	44.5	774.0
57	102	5368	695	528.7	0	421.8	43.6	770.6
	181	5447	672	512.7	0	416.2	42.9	768.0
	1	5448	693	512.8	0	415.9	42.9	768.0
58	102	5549	663	494.4	0	409.4	42.0	764.7
	184	5631	644	481.1	0	404.5	41.4	762.1
	1	5632	1478	480.5	0	404.9	41.4	762.0
59	102	5733	880	348.6	0	399.1	40.6	758.9
	152	5783	835	321.4	0	396.5	40.3	757.4
	1	5784	882	319.7	0	397.0	40.3	757.0
60	32	5815	857	290.0	0	395.4	40.1	756.1
	60	5843	832	274.1	0	394.0	39.9	755.2

Index

E

F

G

H

I

L

M

N

O

P

R

S

T

U